Springer Series on Touch and Haptic Systems

More information about this series at http://www.springer.com/series/8786

Massimiliano Di Luca
Editor

Multisensory Softness

Perceived Compliance from Multiple
Sources of Information

Springer

Editor
Massimiliano Di Luca
School of Psychology, Centre for
    Computational Neuroscience and
    Cognitive Robotics
University of Birmingham
Birmingham
UK

ISSN 2192-2977   ISSN 2192-2985   (electronic)
ISBN 978-1-4471-7061-7   ISBN 978-1-4471-6533-0   (eBook)
DOI 10.1007/978-1-4471-6533-0

Springer London Heidelberg New York Dordrecht

Printed on acid-free paper

Springer is part of Springer Science+Business Media (www.springer.com)

Series Editors' Foreword

This is the 11th volume of *Springer Series on Touch and Haptic Systems*, which is published in collaboration between **Springer** and the **EuroHaptics Society**.

Multisensory Softness is devoted to the perception of object deformation when explored manually. Many research challenges revolve around the topic of multisensory softness perception, such as the medical palpation of internal organs, computational mechanisms of human softness perception or the artificial reproduction of compliant haptic interaction in Virtual Reality. The complexity of haptic and multisensory stiffness perception is due to the interaction between sensory information derived from the skin, the muscles and the tendons, among others, all of which play an important role in properly perceiving material features of objects through the sense of touch. Moreover, soft or hard object properties are good examples of multimodal perception where vision as well as touch plays an important role.

Twenty-six well-known researchers from the field of haptics have contributed to this volume edited by Massimiliano Di Luca. The book is organized into 12 chapters that are grouped into three parts: *Perceptual Softness*, *Sensorimotor Softness*, and *Artificial Softness*. This organization allows the reader to gradually advance in the most relevant topics related to the human perception of softness and to learn about the current state-of-the-art in this field.

June 2014

Manuel Ferre
Marc O. Ernst
Alan Wing

Contents

Part III   Artificial Softness

Contributors

Federico Avanzini Department of Information Engineering, University of Padova, Padova, Italy

Gabriel Baud-Bovy Department of Robotics, Brain and Cognitive Sciences (RBCS), Istituto Italiano di Tecnologia, Genoa, Italy; Faculty of Psychology, San Raffaele Vita-Salute University, Milan, Italy

Wouter M. Bergmann Tiest Faculty of Human Movement Sciences, VU University Amsterdam, Amsterdam, The Netherlands

Matteo Bianchi Department of Advanced Robotics (ADVR), Istituto Italiano di Tecnologia, Genoa, Italy; Research Centre "E. Piaggio", Università di Pisa, Pisa, Italy

Antonio Bicchi Department of Advanced Robotics (ADVR), Istituto Italiano di Tecnologia, Genoa, Italy; Research Centre "E. Piaggio", Università di Pisa, Pisa, Italy

Seungmoon Choi POSTECH, Pohang, Gyungbuk, South Korea

Massimiliano Di Luca School of Psychology, Centre for Computational Neuroscience and Cognitive Robotics, University of Birmingham, Birmingham, UK

Knut Drewing Institute for Psychology, Giessen University, Giessen, Germany

Marc O. Ernst Cognitive Neuroscience Department and Cognitive Interaction Technology-Excellence Cluster, University of Bielefeld, Bielefeld, Germany

Bruno L. Giordano Institute of Neuroscience and Psychology, University of Glasgow, Glasgow, UK

Netta Gurari Department of Robotics, Brain and Cognitive Sciences (RBCS), Istituto Italiano di Tecnologia, Genoa, Italy; Faculty of Psychology, San Raffaele Vita-Salute University, Milan, Italy

Matthias Harders University of Innsbruck, Innsbruck, Austria

Sandra Hirche Institute for Information-Oriented Control, Technische Universität München, Munich, Germany

Seokhee Jeon Kyung Hee University, Seoul, Gyeonggi-do, South Korea

Astrid M.L. Kappers Faculty of Human Movement Sciences, VU University Amsterdam, Amsterdam, The Netherlands

Amir Karniel Department of Biomedical Engineering, Ben-Gurion University of the Negev, Beer-Sheva, Israel

Roberta L. Klatzky Department of Psychology, Carnegie Mellon University, Pittsburgh, PA, USA

Raz Leib Department of Biomedical Engineering, Ben-Gurion University of the Negev, Beer-Sheva, Israel

Amit Milstein Department of Biomedical Engineering, Ben-Gurion University of the Negev, Beer-Sheva, Israel

Ilana Nisky Department of Mechanical Engineering, Stanford University, Stanford, CA, USA

Shogo Okamoto Nagoya University, Nagoya, Japan

Allison M. Okamura Stanford University, Stanford, CA, USA

Markus Rank Research Centre for Computational Neuroscience and Cognitive Robotics (CNCR), University of Birmingham, Edgbaston, UK

Enzo Pasquale Scilingo Research Centre "E. Piaggio", Università di Pisa, Pisa, Italy

Alessandro Serio Research Centre "E. Piaggio", Università di Pisa, Pisa, Italy

Yon Visell Drexel University, Philadelphia, PA, USA

Bing Wu College of Technology & Innovation, Arizona State University, Mesa, AZ, USA

Introduction

One, Few, Many Softnesses

Softness is the subjective impression of the physical compressibility and deformability of objects. It is related to the way objects feel when we touch them, it is a dimension that spans between two extremes: soft, compliant, deformable on the one end and hard, stiff, rigid on the other.

Softness comes readily to mind when handling fabrics and touching pillows, but the dimension soft versus hard does not apply only to objects whose surfaces are deformable. Objects with rigid surfaces (i.e. that are inherently hard to the touch) can be deformable as well—their shape can change when we apply a force to them. For example, switches and buttons can be pressed and released giving the impression of being more or less hard to push. This sensation is to some degree independent of the softness of the material the button is made of.

In general, all objects that have some elastic properties can be defined as soft or hard, but often the terms cannot be used interchangeably without affecting the meaning. For example we could say that "the automatic door is not hard to push", but it would have another meaning to say "the automatic door is soft to push". The term "soft" often connotes the deformation of the object surfaces sensed through the skin rather than changes in the overall shape of the object.

Most objects we interact with in everyday life exhibit some degree of deformability and elasticity. Material compliance affects our comfort (think of pillows and any other surface we rest on), improves our grip (e.g. tool handles and automotive controls), and can even enhance our actions (e.g., when we jump on trampolines or when we shoot arrows with a bow). Deformable objects can have complex behaviours, but evidently our brain can deal with how such objects respond to perturbation allowing us to interact effectively with them. Such understanding is effortless, rapid, and comes natural; this might be at the core of the reason why we have such a good naïve understanding of softness. It might also be for this reason that we use expressions and wording relating to compressibility, extrapolated from the original meaning, in everyday sentences: we "work very hard", "be hardly something", "have a hard time", "are a working stiff", "have a soft spot", "comply with instructions" or could even be "soft in the head".

Because of this intuitive nature, one could be surprised to know that softness and many other material properties are not "direct" properties; i.e. they cannot be estimated using a sensory signal at one time point. The brain makes use of numerous sources of sensory information that are related to material properties of objects and in doing so it integrates sensory information over time. While some signals are directly related to compressibility, most of the sensory signals carry information either about the amount of resistive force or about the amount of indentation. Our brain has specialised mechanisms to deal with all this dynamic information so to create a coherent perceptual representation of the material properties of objects and adjust motor actions accordingly.

The mechanism that allows the perception of the material properties of deformable objects has recently received much scientific attention. Proprioceptively sensed position and force, visual deformation, skin stretching, skin vibration and contact sounds all contribute to perception, but the question is how these multiple signals are combined into a unique softness sensation. Understanding the perceptual aspects of interactions with deformable objects could be beneficial for an appropriate rendering of those interactions in virtual reality. Rendering, for example, can be enhanced by correctly employing sensory information across several sense modalities such as touch, proprioception, vision and audition.

This book focuses on the cognitive mechanisms underlying the use of multiple sources of information in softness perception. It comprises an overview of several aspects of haptic softness with contributions that have been grouped into three parts: the sensory aspect of softness and its multiple signals, the coupling between perception and action, and finally the design, improvement and use of haptic interfaces for softness rendering.

Part I of the book deals with the sensory components and computational requirements of softness perception. Chapter 1 identifies the physical quantities that are used to measure softness, the psychophysical methods to measure how softness is perceived, and finally the modalities involved in sensing softness. Chapter 2 analyses the experimental evidence and proposes a model of how a softness percept is obtained when information from vision and touch is concurrently present. Chapter 3 analyses what type of vibrotactile information is available when interacting with deformable materials and how humans make use of such information to obtain a softness percept. Chapter 4 reviews the relation between material properties and sounds generated during the interaction and it also includes an analysis of the exploitability of such information in recognizing the object's material. Chapter 5 proposes a time-dependent Bayesian model that can account for the way the brain integrates multiple sensory signals with previous knowledge of what the material of the object is.

Material properties modulate impedance, whose effect can only be sensed dynamically, i.e., while indenting and perturbing the objects. Motor actions are therefore important components in the perception of material properties. In compliance perception, the type of exploratory procedures employed and the pattern of movement performed are very important in determining the precision of the estimate. Hence, Part II of the book deals with these motor components of the interaction with soft objects. Here, Chap. 6 investigates how the acquisition of sensory information can be improved through the use of different contact areas and different movement parameters. Chapter 7 presents experimental data about the errors involved in finding the centre of a force field generated by a force-feedback device. Chapter 8 proposes a dynamic state model of interaction with a haptic interface for the detection of delay in force generated in response to participant movement. Chapter 9 covers the experimental findings on how softness perception is affected by the manipulation of the time delay between movement and force feedback, considering the prediction of several perceptual models.

Part III of the book focuses on the identification of exploitable guidelines to help replicate softness in artificial environment. Chapter 10 investigates methods to provide artificial sensory signals to enhance or substitute the sensory feedback obtained during the interaction with compliant objects. Chapter 11 describes a display that can mimic the direct tactile contact with objects of different compliances by modulating the tension of a stretchable fabric. Chapter 12 introduces a technique to change the perceived softness of physical objects by using a haptic device that overlays forces to the resistive force generated by the object during interaction.

The three parts of the book cover how sensory signals are used to perceive softness, how humans interact with compliant objects and how to reproduce interaction with soft virtual objects. Overall, the book offers a multidisciplinary analysis of how sensory signals from different modalities (skin pressure, proprioception, tactile vibration, vision and audition) are combined so to lead to the impression of material compliance—the perception of softness.

Acknowledgments

I thank every chapter author and every contributor. In particular, I am grateful to Federico Fontana, Laurie Heller, Hiroyuki Kajimoto, Hoi Fei Kwok, Stephen McAdams, Chris Miall, Darren Rhodes, Alan Wing, Jeremy Wyatt and Claudio Zito for their comments and discussions. Finally, I would like to acknowledge the great work of Charlie Bambford and Luke Rogerson who helped in organizing the reviews and thoroughly copy editing the contributions.

Part I
Perceptual Softness

Chapter 1
Physical Aspects of Softness Perception

Wouter M. Bergmann Tiest and Astrid M.L. Kappers

1.1 Introduction

This chapter discusses the relationship between physical and perceived softness and hardness. In order to properly study softness perception, it is important to specify what softness is, and how it is measured. Section 1.2 of this chapter deals with physical, objective scales for softness and hardness. Methods are then outlined in Sect. 1.3 to accurately characterise the human ability for perceiving softness. Section 1.4 discusses what cues are available for softness perception, and how they are used. Finally, this chapter describes how the different cues work together in various situations.

1.2 Physical Measures of Softness

A material's softness relates to its ability to deform under pressure. Deformation can be elastic, viscous, or otherwise; however in this book, we will be mostly concerned with elastic deformation. In physics and engineering, several measures are available to describe the material's softness. The most basic measure is an object's *stiffness*, which is the ratio of the force applied to the object and the amount of resulting deformation in the direction of the applied force. It is equivalent to the spring constant of a linear spring, which is usually indicated by the symbol k:

$$F = k \cdot \Delta l, \tag{1.1}$$

W.M. Bergmann Tiest (✉) · A.M.L. Kappers
Faculty of Human Movement Sciences, VU University Amsterdam,
Amsterdam, The Netherlands
e-mail: W.M.BergmannTiest@vu.nl

© Springer-Verlag London 2014
M. Di Luca (ed.), *Multisensory Softness*, Springer Series on Touch and Haptic Systems,
DOI 10.1007/978-1-4471-6533-0_1

where F is the force and Δl is the change in length. This relation is known as *Hooke's law*. The unit of stiffness is N/m. The same measure can also be expressed as a *compliance*, which is the inverse of stiffness, with units of m/N.

A careful distinction should be made between an idealised, zero-length, one-dimensional linear spring, and a real compliant object. The former can be used conveniently to model simple elastic behaviour, but is often not representative of a real-life situation. It should be noted that a real object's compliance (or stiffness) need not be a single value, and can depend on the direction or magnitude of the force. Furthermore, the stiffness will depend on the object's dimensions: the thicker the object, the larger the deformation will be for a given force. The wider the object, with the force spread out over a larger area, the smaller the deformation will be. A measure that takes these aspects into account, at least for so-called 'linear materials' that behave according to Hooke's law, is the *elastic modulus* or *Young's modulus*. This is defined as the ratio of *stress* and *strain*. Referring to Fig. 1.1, stress is the applied force per unit area F/A, while strain is the relative change in length in one direction: $\Delta l / l$. For the elastic modulus we therefore obtain

$$E = \frac{F/A}{\Delta l / l} = \frac{Fl}{A \Delta l}, \tag{1.2}$$

in units of N/m^2 or Pa. Because it is based on *relative* deformation and force *density* (i.e., pressure), it is less dependent on the object's dimensions. In that respect, it can be said that stiffness is a property of a specific object, whereas the elastic modulus is a property of the material.

Another physical measure for hardness is the *Shore durometer scale*. This measure is based on the indentation a probe makes when pressed into the material with a given force, and can be measured using a durometer. The Shore system includes a set of scales corresponding to different probe sizes and shapes (e.g. conical or spherical) and indentation forces. The hardness is then expressed as a number between 0 and 100 on that specific scale, corresponding to an indentation between 100 and 0 mil

Fig. 1.1 Quantities involved in the deformation of an object: F is the applied force; A is the area over which it is applied; l is the original thickness of the object; Δl is the change in thickness due to the application of the force

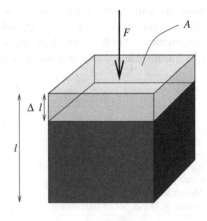

(1/1,000th of an inch), respectively. Similar systems include the Vickers hardness test and the Rockwell hardness test. These scales are mainly intended for testing relatively hard materials such as steel, and for this reason they are not used very often in the context of human perception.

1.3 Psychophysical Measurement of Softness Perception

Several psychophysical techniques are available for characterising softness perception. The most important ones are *magnitude estimation, discrimination measurements, matching,* and *identification*. These will be discussed in this section.

1.3.1 Magnitude Estimation

Using magnitude estimation, the physical magnitude of a stimulus can be linked to its perceived magnitude. In general, participants are presented with a stimulus and asked to express their perception of its magnitude as a number (e.g. on a scale from 0 to 100) or as a position on a line (visual analogue scale, VAS). Sometimes a standard is used, to which a fixed number is assigned. When a test stimulus is presented that feels twice as hard as the standard, the participant should respond with a number that is twice as high, and so on. In other situations, the participant is completely free to choose any scale. When used with a wide range of compliant stimuli, magnitude estimation can provide insight into the functional relationship between physical and perceived softness.

In an experiment where participants squeezed different types of rubber, it was found that this relationship was nonlinear (Harper and Stevens 1964). The curve relating physical stiffness to subjective hardness ratings could be fitted by a power law with an exponent of 0.8. This value being less than 1 indicates that the steepness of the curve levels off at greater hardness. The inverse relationship, with an exponent of −0.8, was found for perceived softness ratings, indicating that hardness and softness are perceptual opposites. Later research, however, showed that the exponent can vary substantially across participants (Nicholson et al. 2000). In this experiment, participants pressed down with both hands on a mechanical device and had to rate its stiffness. Fitted exponents ranged from 0.92 to 2.56, but were quite consistent within participants between two sessions that were at least two weeks apart (correlation coefficient $R = 0.84$). Furthermore, an experiment with rendered virtual springs found a power law exponent of 0.93 (Varadharajan et al. 2008). Note that in these experiments, there was no direct skin contact with a compliant material. It has been shown that this has an effect on the perceived magnitude of softness (Friedman et al. 2008). In that experiment, participants judged the softness of silicone rubber disks in various conditions, with or without direct skin contact. In particular the stimuli at the harder end of the range were judged to be softer when touched using a tool, compared to direct skin contact. Also, hard stimuli that were touched passively (pressed down on the skin) were judged to be softer than the same stimuli touched actively. In short, magnitude estimation of softness varies from person to person, and also with the availability of direct skin contact and with the mode of touch (active or passive).

1.3.2 Discrimination

In perceptual discrimination experiments, participants are repeatedly presented with two stimuli that differ in magnitude. They then have to indicate which stimulus has the higher magnitude (two-alternative forced-choice task, 2AFC). There are different ways of obtaining a discrimination threshold through a 2AFC task. The *method of constant stimuli* defines a set of test stimulus magnitudes beforehand, usually spread around a fixed reference stimulus (for example 8 test magnitudes; 4 below and 4 above the reference magnitude). Each pair of stimuli is presented a number of times (e.g. 10) in random order. When the fraction of times each test stimulus is chosen as the one with the higher magnitude is plotted against the test stimulus magnitude, a *psychometric curve* is created. A suitable function can be fitted to this curve, for instance one based on the cumulative Gaussian distribution:

$$f(x) = \tfrac{1}{2} + \tfrac{1}{2}\mathrm{erf}\left(\frac{x - x_{\mathrm{ref}}}{\sqrt{2}\sigma}\right) \qquad (1.3)$$

Here, x is the magnitude of the test stimuli, x_{ref} is the magnitude of the reference stimulus, and the fit parameter σ is a measure for the steepness of the curve, and corresponds to the difference between the 0.84 level and the 0.5 (chance) level of the curve. This threshold σ is termed the Just Noticeable Difference (JND). The lower this value, the more precise the stimulus continuum can be perceived. An example of a psychometric curve with fitted function is shown in Fig. 1.2.

Other expressions for psychometric curves have been suggested, such as the cumulative logistic distribution or the Weibull distribution. Furthermore, these functions can be modified to include other parameters to account for the *guess rate* (the fraction of correct responses that can be expected purely by chance) and the *lapse rate* (the fraction of incorrect responses present even at a stimulus magnitude far above threshold level, due to lapses in the participant's concentration) (Wichmann and Hill 2001).

Alternatively, a *staircase procedure* can be used to estimate the JND. This procedure also uses a 2AFC task, consisting of presenting a pair of test and reference stimuli on each trial, and asking the participant to indicate the one with the higher magnitude. The test stimulus in the next trial is then determined by the participant's answer: one step more intense when the reference stimulus was chosen a certain number of times in a row, or one step less intense when the test stimulus was chosen. After a certain number of trials or a certain number of reversals of the direction of the staircase, the procedure is terminated. Then, the difference between the test

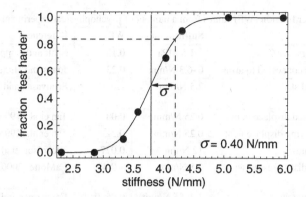

Fig. 1.2 Data from a hardness discrimination experiment. The *dots* indicate the fraction of times that a test stimulus was found to be harder than the reference stimulus, which is indicated by the *vertical line*. The *solid curve* is a fit to the data. The *dashed line* indicates the 0.84 level. The difference between the corresponding stiffness value and the reference value is the discrimination threshold σ, as indicated

and reference stimulus, averaged over the last, say, 20 trials, is representative for the JND. The exact proportion on the psychometric curve to which this difference corresponds, depends on the details of the staircase procedure. For instance, moving one step forward after three correct responses in a row, and one step back after a single incorrect response (a 3-up-1-down staircase) will converge at the 0.79 level. In general, the staircase procedure requires considerably fewer trials than the constant stimuli method, but can be less precise and might not always converge.

In addition to the threshold value σ, discrimination performance can also be expressed as the ratio of threshold and reference magnitude, which is called the *Weber fraction*, $W = \sigma/x_{\text{ref}}$. By determining such thresholds for a range of reference magnitudes, and under various perceptual conditions, the relative importance of different sources of information can be assessed. Often, the Weber fraction is more or less constant over a considerable range of stimulus magnitudes. This phenomenon is known as the *Weber-Fechner law*, which states that the Just Noticeable Difference (JND) is a constant fraction of the stimulus magnitude.

One of the earliest softness discrimination measurements yielded a Weber fraction of about 0.13 (Scott and Coppen 1939). This was based on a comparison of different rubber cylinders with elastic moduli of around 1.5 MPa. Later, using a mechanical device, where participants did not touch a deformable surface, Weber fractions for stiffness discrimination were found to be around 0.22 (Tan et al. 1995). This illustrates the importance of the information from the surface deformation, which will be discussed below. When participants were allowed to base their judgement on the force at the end of the displacement, the average Weber fraction was as low as 0.08, but this can be attributed to force discrimination rather than stiffness discrimination. Weber fraction values for stiffness discrimination have been found to be replicable, with an average difference between test and re-test of less than 2 % (Nicholson et al. 1997).

Table 1.1 Weber fraction (WF) values from a number of psychophysical experiments

stimuli	Stiffness range	WF	References
Rubber cylinders	~1.5 MPa	0.13	Scott and Coppen (1939)
Electric motors connected to arms	0–6.3 N/mm	0.23	Jones and Hunter (1990)
Silicon rubber disks	2.5 N/mm	<0.4[a]	Srinivasan and LaMotte (1995)
Rigid plates, fixed displacement	0.25 N/mm	0.08	Tan et al. (1995)
Rigid plates, roving displacement	0.25 N/mm	0.22	Tan et al. (1995)
Single rigid plate	12 N/mm	0.077	Nicholson et al. (1997)
Silicon rubber disks pressed with stylus	2.5 N/mm	<0.4[a]	LaMotte (2000)
Silicon rubber blocks	1.2–55 N/mm	0.25–0.29	Freyberger and Färber (2006b)
Silicon rubber cylinders	4.8–24 N/mm	0.15	Bergmann Tiest and Kappers (2009)
Rigid plates	10–36 N/mm	0.5	Bergmann Tiest and Kappers (2009)
Virtual cube explored with rigid probes	0.63 N/mm	0.29	Kuschel et al. (2010)

[a]Estimate; no discrimination threshold is reported, only % correct

Turning back to deformable materials, it has been found that softness discrimination sensitivity was better when participants pressed down on the stimuli than when they pinched them between thumb and index finger (Freyberger and Färber 2006a). With participants pinching the stimuli, there was no difference found in discrimination performance between 20-mm-thick and 40-mm-thick stimuli, when their hardness was expressed as a stiffness value (Weber fractions around 0.15) (Bergmann Tiest and Kappers 2009). To conclude, softness discrimination is possible with a precision of around 13–15 %, and depends on how the stimulus is touched. These discrimination thresholds are summarised in Table 1.1.

1.3.3 Matching

In order to compare perception under different conditions, a matching experiment can be performed. In such an experiment, the magnitudes that are perceived to be equal are determined between two different types of stimuli. This is called the Point of Subjective Equality (PSE). This is done by allowing the participant to adjust the magnitude of a test stimulus until it perceptually matches that of the reference stimulus.

Another method to determine this point is through a discrimination experiment, as described above. Using the method of constant stimuli and a 2AFC task, a psychometric curve can be measured. To this curve, a function of the same form as Eq. (1.3) can be fitted, but with x_{ref} as a free parameter. This fit parameter then corresponds to the PSE.

Similarly, a 1-up-1-down staircase procedure with a 2AFC task can be used effectively to estimate the PSE. After a number of trials, this procedure converges at the point where the test and reference stimulus are of equal perceived magnitude. In order to avoid a bias, it is recommended to use two interleaved staircase procedures: one starting from above and one from below the reference magnitude. In this way, the experiment 'zooms in' on the PSE.

An experiment involving matching between the two hands using electrical motors yielded a linear relationship between reference stiffness and matching stiffness, with $R = 0.97$ (Jones and Hunter 1990). From the spread in the participant's responses in repeated trials, a discrimination threshold could also be calculated, which corresponded to a Weber fraction of 0.23. This is very similar to the value of 0.22 found using a discrimination experiment (Tan et al. 1995), showing that different methods yield very comparable results.

The main point of such matching experiments, however, is to compare perception under different circumstances. For example, a study investigated the influence of stimulus location on stiffness perception using a discrimination experiment with virtual springs (Wu et al. 1999). In the haptics-only condition, the spring furthest away from the participant was perceived to be about 10 % softer than the spring closest to the participant. This bias disappeared when visual feedback was provided, and also the discrimination threshold improved somewhat. This indicates that haptic perception is dependent on location, but visual feedback can be used to compensate for biases that arise as a result.

Matching experiments can also be used with different stimulus types. In a softness matching experiment between thick (40 mm) and thin (20 mm) silicon rubber stimuli, it was found that the matched stimuli did not differ significantly in terms of Young's modulus, but the difference was significant when expressed as a stiffness value (Bergmann Tiest and Kappers 2009). This suggests that people either use stiffness information and compensate for differences in object geometry, or disregard the stiffness altogether and only pay attention to the surface deformation.

1.3.4 Identification

Lastly, absolute identification of softness values is of importance for doctors and veterinarians in the practice of palpation. This involves judging the softness of body parts, for example for the diagnosis of breast cancer or determining the state of pregnancy of cows. To determine human performance in absolute identification of softness, a number of different softness values are presented and the participant is asked to categorise them. In a study using virtual stiffness, it was found that trained veterinarians were almost twice as good in identifying five different values of stiffness than novice veterinary students (Forrest et al. 2011). This indicates that stiffness identification can be learned.

1.4 Cues for the Perception of Softness

This section focuses on the different types of information that are available for basing a softness judgement on, and how these are combined to form a single percept. Important sources of information are (1) the ratio between force and displacement, and (2) the shape and size of the deformation of the surface. Although softness is primarily a haptic property, visual as well as tactual cues can be used to obtain this information.

1.4.1 Visual Cues

Visual cues can convey information about the displacement, which can be used in combination with haptic force cues for softness perception. The importance of this role of visual information was shown in an experiment where participants performed a stiffness discrimination task using an apparatus that presented virtual springs (Srinivasan et al. 1996). At the same time, they were shown the deformation of the springs on a computer screen. The results showed that when there was a mismatch between the visual and haptic information, participants based their judgement, for the most part, on the visual information. They felt a physically stiffer spring as less stiff when it showed a greater deformation on the screen. This illusion was used to simulate haptic feedback in a situation where participants pushed against a passive handle, and were shown the deformation of a spring on a screen (Lécuyer et al. 2000). The actual displacement was very small, and the physical force/displacement relationship was the same for all stimuli, so displacement information could only be obtained visually. Even so, participants reported feeling differences between the different springs and could discriminate stiffness with a Weber fraction of 6 %. When they had to perform stiffness discrimination between a virtual spring (with only visual displacement and haptic force information) and a real spring (with haptic force and displacement information), the Weber fraction was found to be 13 %. Although the illusion is very compelling, there is a limitation to the amount of mismatch between haptic and visual information. It was found that when the discrepancy becomes greater than 55 %, a difference between haptic and visual information becomes perceivable (Kuschel et al. 2008).

When both visual and tactual cues are present, one would expect discrimination to always be better than when either cue is available by itself. In a study involving discrimination based on either visual, proprioceptive, or combined cues, however, it was found that some participants did better in one condition, and others in another (Gurari et al. 2009). On average, discrimination in the combined condition, with a Weber fraction (WF) of 0.039, was worse than in the proprioceptive-only condition (WF 0.036), but better than in the visual-only condition (WF 0.056). The findings suggest that some people are more visually oriented, whereas others are more touch-oriented.

The studies cited above involved visuohaptic perception of the stiffness of springs. In that case, only the displacement of the spring can be perceived visually and the force can only be perceived haptically. When a deformable object is used, however, instead of a spring, the surface deformation cue becomes available, which can be perceived visually. This was demonstrated in an experiment where participants performed magnitude estimation of softness of silicon rubber cylinders in haptic, visual, and visuohaptic conditions (Drewing et al. 2009). In the haptic and visuohaptic conditions, they explored the stimuli with or without a blindfold, whereas in the visual condition, they watched another person explore the stimuli. In all conditions, participants were able to distinguish between the stimuli, but in the haptic condition, they were judged somewhat harder than in the visual condition, with the visuohaptic judgements in between (on average a little closer to the visual than to the haptic judgement). From this experiment, it can be concluded that visual cues alone are sufficient for softness perception, and that when both visual and haptic cues are available, the judgement is based for 55 % on visual and 45 % on haptic information.

Kuschel et al. (2010) carried out an experiment to study whether a judgement is arrived at by either first combining position and force information in each modality separately and then integrating the softness information from both modalities, or by first integrating the haptic and visual position information separately from the haptic and visual force information, and then combining the two. In one condition both force and position information was available to both modalities, and in another condition only visual position and haptic force information was available. The authors' assumption was that if information is first integrated and then combined, the precision of perception would be about the same in both conditions, since the haptic position and visual force information are considered relatively unreliable and do not contribute much to discrimination performance. On the other hand, if information is first combined and then integrated, essential information is missing and discrimination performance is much worse in the second condition. Since discrimination in the first condition was found to be much better than in the second, the authors concluded that a softness percept is first formed in each modality separately, and only then integrated. Furthermore, based on a second experiment, they concluded that the weight factors attributed to the two modalities are close to optimal when haptic and visual information is congruent, but they remain the same when the information is incongruent, and therefore are not optimal in that situation. Further discussion and experimental results about visuohaptic perception of softness are covered in Chap. 2. Chapter 5 analyses the computational requirements involved in integrating force and position information.

1.4.2 Tactual Cues

Tactual sources of information include the ratio between force and displacement of the fingers, and the force distribution over the contact area, as well as its shape and size; see Fig. 1.3. The possible contribution of vibrotactile information is analysed

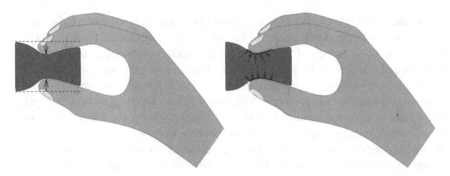

Fig. 1.3 Sources of information about an object's softness. *Left* Amount of linear displacement of the fingers when a force is applied. *Right* Shape and size of the contact area, and force distribution over this area

in Chap. 3. Tactual cues can be further subdivided into kinaesthetic cues (relating to the perception of forces on, and movement of, limbs or fingers) and cutaneous cues (relating to the perception of pressure on the skin). Displacement of the fingers is mostly a kinaesthetic sensation, whereas force information can be sensed both kinaesthetically and cutaneously. On the other hand, the force distribution and the contact area can only be sensed cutaneously. Using local anaesthesia, it has been shown that cutaneous information is both necessary and sufficient for accurate softness perception (Srinivasan and LaMotte 1995). Thus, with only kinaesthetic information, participants could not discriminate between rubber disks that differed by a factor of 1.9 in stiffness (recall the Weber fraction of about 0.15 for normal softness perception, corresponding to a difference of a factor of 1.15). In another experiment from the same article, passive touch was used, which involved pressing the stimuli down onto the participants' fingers. Here only cutaneous information was available and participants were able to use the cutaneous force information and the surface deformation to discriminate between the disks. However, with spring cells having rigid surfaces, this was not possible, and the addition of kinaesthetic displacement information was necessary for discrimination. Furthermore, experiments have shown that softness discrimination is possible using a tool (a stylus), which conducts only force and displacement information (LaMotte 2000). Also with the tool, cutaneous cues are sufficient for force sensation, but kinaesthetic cues are necessary for the displacement.

The available cutaneous cues include the contact area spread rate (CASR) (Bicchi et al. 2000). This is the slope of the function relating the normal force and the size of the contact area. With this cue present in a mechanical softness display (see Chap. 11), participants were 30 % more precise in softness discrimination compared to a force/displacement display without the CASR cue. When these two cue types are combined, discrimination is even better, but still not as precise as direct interaction with a real deformable surface (Scilingo et al. 2010). To precisely pinpoint the importance of this direct interaction, Weber fractions from a discrimination experiment with silicon rubber cylinders were compared to those obtained with the same cylinders, but with rigid steel disks interposed between the objects' surfaces and the participants' fingers (Bergmann Tiest and Kappers 2009). In the latter case, participants could still use cutaneous cues but only for force information, since there was no deformation of the surface. Weber fractions went up to 0.5, highlighting the importance of the surface deformation cue. Under the assumption that stiffness cues and surface deformation cues are combined in a statistically optimal fashion, this suggests that about 90 % of the information is obtained from the surface deformation.

In order to further study the role of these cues, an experiment was set up in which the stiffness and surface deformation cues were decoupled (Bergmann Tiest and Kappers 2009). This was done by means of a stimulus set with a sandwich configuration: Between the top and bottom silicon rubber layers, an incompressible part was inserted, while keeping the total thickness of the stimuli the same as that of the stimuli consisting of silicon rubber all the way through. In this way, the surface characteristics and finger span were kept the same, but the relation between stiffness and Young's modulus differed between the two stimulus sets. A matching experiment determined which values were required for participants to feel the two stimulus types to be equal in softness (PSE). The results are shown in Fig. 1.4, where the relationship between Young's modulus and stiffness is plotted for the two types of stimuli (dashed for the homogeneous type and dotted for the sandwich type). In the figure, arrows indicate the average softness matches made between the two types of stimuli. The grey lines are a fit to the direction of these arrows, showing possible curves of equal perceived softness. As can be seen, the lines for the softer stimuli (close to the bottom left of the graph) are oriented much more towards the horizontal

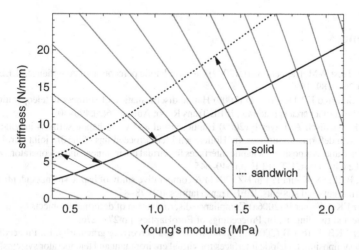

Fig. 1.4 Relationship between the object stiffness and the material's Young's modulus for two types of stimuli (*dashed and dotted curves*). The *arrows* indicate the matches made between these stimulus types. The *grey lines* illustrate possible curves of equal perceived softness, based on the matches. Based on data from Bergmann Tiest and Kappers (2009)

than for the harder stimuli (top right of the graph). This suggests that participants' attention shifts towards stiffness for softer stimuli, and more towards the Young's modulus for harder stimuli. Since the Young's modulus is a property of the material, whereas stiffness is a property of the object, we can say that for softer stimuli, the determining factor is how the whole object can be compressed, whereas for harder stimuli, it is how the material in contact with the fingers deforms under pressure.

1.5 Conclusions

In conclusion, it can be said that the relationship between physical and perceived softness has been studied quite extensively. Physical softness is captured by the stiffness (or compliance) and the elastic modulus. Since the stiffness can be perceived both haptically and visually by using available force and displacement components, it is of greater importance when the modalities are used together. However, when only haptic information is used, and especially for stiffer objects, the elastic modulus becomes the most important factor. Since this component of softness is sensed through the cutaneous channel within the haptic modality, it is of the utmost importance to appropriately stimulate this channel when attempting to artificially display the sensation of softness.

Acknowledgments This work has been partially supported by the European Commission with the Collaborative Project no. 248587, "THE Hand Embodied", within the FP7-ICT-2009-4-2-1 program "Cognitive Systems and Robotics".

References

Bergmann Tiest WM, Kappers AML (2009) Cues for haptic perception of compliance. IEEE Trans Haptics 2(4):189–199

Bicchi A, Scilingo EP, De Rossi D (2000) Haptic discrimination of softness in teleoperation: the role of the contact area spread rate. IEEE Trans Robot Autom 16(5):496–504

Drewing K, Ramisch A, Bayer F (2009) Haptic, visual and visuo-haptic softness judgments for objects with deformable surfaces. In: Hollerbach J (ed) Proceedings of the 3rd joint EuroHaptics conference and symposium on haptic interfaces for virtual environment and teleoperator systems, Salt Lake City. USA, UT, IEEE, pp 640–645

Forrest N, Baillie S, Kalita P, Tan HZ (2011) A comparative study of haptic stiffness identification by veterinarians and students. IEEE Trans Haptics 4(2):78–87

Freyberger FKB, Färber B (2006a) Compliance discrimination of deformable objects by squeezing with one and two fingers. In: Proceedings of EuroHaptics, pp 271–276.

Freyberger FKB, Färber B (2006b) Psychophysics and perceiving granularity. In: Proceedings of the 14th symposium on haptic interfaces for virtual environment and teleoperator systems. IEEE, pp 387–393.

Friedman RM, Hester KD, Green BG, LaMotte RH (2008) Magnitude estimation of softness. Exp Brain Res 191(2):133–142

Gurari N, Kuchenbecker KJ, Okamura AM (2009) Stiffness discrimination with visual and proprioceptive cues. In: Hollerbach J (ed) Proceedings of the 3rd joint EuroHaptics conference and symposium on haptic interfaces for virtual environment and teleoperator systems, Salt Lake City. USA, UT, IEEE, pp 121–126

Harper R, Stevens SS (1964) Subjective hardness of compliant materials. Q J Exp Psychol 16: 204-215.

Jones LA, Hunter IW (1990) A perceptual analysis of stiffness. Exp Brain Res 79(1):150–156

Kuschel M, Freyberger F, Färber B, Buss M (2008) Visual-haptic perception of compliant objects in artificially generated environments. Visual Comput 24(10):923–931

Kuschel M, Di Luca M, Buss M, Klatzky RL (2010) Combination and integration in the perception of visual-haptic compliance information. IEEE Trans Haptics 3(4):234–244

LaMotte RH (2000) Softness discrimination with a tool. J Neurophysiol 83(4):1777–1786

Lécuyer A, Coquillart S, Kheddar A, Richard P, Coiffet P (2000) Pseudo-haptic feedback: can isometric input devices simulate force feedback? In: Proceedings of the virtual reality annual international symposium, pp 83–90.

Nicholson L, Adams R, Maher C (1997) Reliability of a discrimination measure for judgements of non-biological stiffness. Manual Ther 2(3):150–156

Nicholson L, Adams R, Maher C (2000) Magnitude estimation of manually assessed elastic stiffness: stability of the exponent. Percept Mot Skills 91:581–592

Scilingo EP, Bianchi M, Grioli G, Bicchi A (2010) Rendering softness: integration of kinaesthetic and cutaneous information in haptic devices. IEEE Trans Haptics 3(2):109–118

Scott Blair GW, Coppen FMV (1939) The subjective judgements of the elastic and plastic properties of soft bodies; the "differential thresholds" for viscosities and compression moduli. Proc R Soc Lond Ser B Biol Sci 128(850):109–125

Srinivasan MA, LaMotte RH (1995) Tactual discrimination of softness. J Neurophysiol 73(1): 88-101.

Srinivasan MA, Beauregard GL, Brock DL (1996) The impact of visual information on the haptic perception of stiffness in virtual environments. Proc ASME Dyn Syst Control Div 58:555–559

Tan HZ, Durlach NI, Beauregard GL, Srinivasan MA (1995) Manual discrimination of compliance using active pinch grasp: the roles of force and work cues. Percept Psychophysics 57(4):495–510

Varadharajan V, Klatzky R, Unger B, Swendsen R, Hollis R (2008) Haptic rendering and psychophysical evaluation of a virtual three-dimensional helical spring. proc. In: Symposium on haptic interfaces for virtual environments and teleoperator systems. IEEE, pp 57–64.

Wichmann FA, Hill NJ (2001) The psychometric function: I. Fitting, sampling, and goodness of fit. Percept Psychophysics 63(8):1293–1313.

Wu WC, Basdogan C, Srinivasan MA (1999) Visual, haptic, and bimodal perception of size and stiffness in virtual environments. Proc ASME Dyn Syst Control Div 67:19–27

Chapter 2
Visual-Haptic Compliance Perception

Roberta L. Klatzky and Bing Wu

2.1 Analysis of Softness as a Higher-Order Property

Traditional psychophysical methods are so-called because some physical dimension is carefully manipulated while people make judgments of its psychological impact (see Chap. 1). Typical psychophysical measures are: the detection threshold (minimum stimulus intensity required for conscious perception), the discrimination threshold or Just Noticeable Difference (JND; minimum difference in intensity required for discrimination), and parameters of the function relating perceived intensity to physical intensity across a range of values on the target dimension. Early attempts to measure these psychophysical variables tended to focus on univariate quantities. Thus, for example, one can find values for the threshold intensity of physical dimensions such as length, brightness, or weight, as well as other quantitative dimensions with less obvious physical interpretations such as salt dilution or voltage of a current applied to the skin (see Woodworth and Schlossberg 1960). One can also find, from so-called *magnitude estimation* tasks, that perceived stimulus magnitude tends to be related to physical signal intensity in the form of a logarithmic function (Fechner 1860) or a power function, the exponent of which provides a summary measure of the perceptual transduction output for a given dimension (Stevens 1975).

The present chapter focuses on stiffness, the mathematical inverse of compliance. Stiffness is inherently a higher-order property, in the sense that it is computed from the relation between two underlying quantities. By Hooke's Law, stiffness (denoted k) is the relation between displacement (d, change in position or length) and the force that produces that displacement (F), as specified by the equation, $F = kd$. Given that stiffness is defined by the ratio of two physical variables, force and posi-

---

R.L. Klatzky (✉)
Department of Psychology, Carnegie Mellon University, Pittsburgh, PA, USA
e-mail: klatzky@cmu.edu

B. Wu
College of Technology & Innovation, Arizona State University, Mesa, AZ, USA

© Springer-Verlag London 2014   17
M. Di Luca (ed.), *Multisensory Softness*, Springer Series on Touch and Haptic Systems,
DOI 10.1007/978-1-4471-6533-0_2

Fig. 2.1 Judging stiffness (k) from perceived force (F) and displacement (d)

Visual

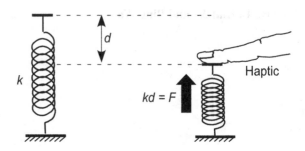

tion, a psychophysical characterisation means examining first how the individual variables and the ratio are perceived. This chapter is concerned with an additional source of complexity, namely, the contributions of vision and haptic perception to the perception of stiffness, as illustrated in Fig. 2.1.

2.1.1 The Components of Stiffness Through Vision and Haptics

In this section we briefly consider how displacement and force are perceived through vision and haptic perception in isolation. There are, then, four cases to consider, resulting from the combination of two properties and two modalities.

The first case is the perception of displacement by the visual modality. Some experiments have shown that visual judgments of length in near space are proportional to physical length, as indicated by power-function exponents of 1.0 (Seizova-Cajic 1998; Teghtsoonian and Teghtsoonian 1965). Linear relations, however, can still show bias; for example, the slope relating judged length to physical length was shown in one experiment to be on the order of 0.9, introducing systematic under-estimation bias (Keyson 2000).

Next, consider the perception of displacement by touch. For haptic judgments of length of a line, power-function exponents extracted from magnitude estimation vary between 0.8 and 1.2, suggesting greater distortion than is generally found for vision. The exponent depends on the stimulus range (a usual phenomenon in scaling experiments), orientation of the stimulus (Lanca and Bryant 1995), and circumstances of exploration. For example, tracing the length of the line with the fingertip leads to relatively low error (Stanley 1966), but the length of a stimulus felt through a pinch grasp is overestimated (Seizova-Cajic 1998; Teghtsoonian and Teghtsoonian 1965, 1970).

Force, the second component of stiffness, is sensed haptically through cutaneous deformation and mechanoreceptors in muscles, tendons, and joints. Characterizations of haptic force perception have been widely variable (see Jones 1986, for review), depending on the range of forces tested and the experimental procedures. For example, an exponent has been reported of 1.7 for hand-gripping forces (Stevens and Mack 1959), 0.8 for lifted weights (Curtis et al. 1968—in which case force is derived from the relation to mass and gravitational acceleration), and 1.0 for tangential forces applied to the fingertip (Paré et al. 2002).

Visual cues to force are minimal and essentially heuristic. Surface deformation patterns offer cues to compressive force (Wang et al. 2001), although generally without metric scaling. An interesting approach was offered by Sun et al. (2008), who analysed camera images of the fingernail and nail bed, then used changes in colour to predict the applied force. Beyond these direct cues, people may exhibit what Bayesian modeling refers to as "priors"—expectations based on past experience (see Chap. 5 for a discussion about Bayesian models). Visually based priors come into play because our experiences of force generally occur in the context of visual experience. It has been suggested that the sensation of force through mechanical interaction and the corresponding displacements perceived by vision become associated in long-term memory, and thus kinematic features in a visual percept can be matched to stored haptic experiences to infer force (White 2012). Michaels and De Vries (1998) asked their participants to judge the force exerted by a videotaped puller who gripped a handle and made pulls to specified target forces without moving their feet. Their results showed high correlations between visual judgments and target forces.

2.1.2 Multi-Modal Perception of Stiffness Components

Even if we were to know, for each modality, the psychophysical functions relating perceived force and displacement to their physical values, we would not necessarily be able to quantitatively predict the perceived stiffness. For one thing, internalized quantities may not behave like ratio scales, so the simplicity of Hooke's Law is questionable here. More importantly, the presence of two modalities tends to lead to perceptual outcomes that differ from the simple effects of either component in isolation. Here we consider interactions between vision and haptics in the estimate of the component properties of stiffness, displacement and force.

Consider first the multisensory integration of spatial cues in displacement perception. Relative displacement might be computed from comparing successive representations of the effector location. In relevant work, it has been found that when a person's hand is localized in space using proprioceptive and visual information, the two type of cues are averaged in a weighted linear manner, so that the reciprocal of the bimodal variance is equal to the sum of the reciprocals of the two unimodal variances (van Beers et al. 1999). Relative displacement can also be computed by comparing simultaneous positions of effectors. In a classic study, Ernst and Banks (2002) demonstrated that the integration of visual and haptic information concerning distance between the fingers follows a Maximum-Likelihood Estimation model (MLE). The perceptual outcome was a weighted linear combination of estimates from each modality, where the weights were inversely related to the estimates' variability. That is, the more reliable estimates received more weight.

Another way to compute displacement is to compare the position of a surface indented by an effector to a surrounding surface. One of our studies found that visual and haptic cues combine in a linear fashion in the perception of surface displacement (Wu et al. 2008). Participants were asked to indent a probe into a soft surface until it "bottomed out" against a barrier; they then attempted to estimate the extent of the indentation. Visual cues were present from the indentation per se and also from deformation of a grid pattern painted on the surface. Haptic feedback was experimentally manipulated by varying stiffness of an underlying membrane. We found that resisting forces arising from surface indentation heightened the perception of deformation. For a constant physical indentation, higher force led to greater perceived surface indentation. A regression analysis found that the data could be described by a weighted linear combination of visual and haptic cues, although the optimality of the weighting was not tested.

We next turn from multi-modal perception of displacement to multi-modal perception of force. The haptic perception of force direction, for example, can be significantly enhanced with congruent visual cues. Bargagli et al. (2006) found that the threshold of force-direction discrimination was reduced from 25.6° in the haptic-only condition to 18.4° when congruent visual cues were introduced, but increased to 31.9° when the haptic and visual inputs were incongruent. The size/weight illusion (large objects are perceived as lighter than smaller objects of equal weight) might be the most extensively studied phenomenon with respect to visual influence on the perception of force magnitude. Ellis and Lederman (1993) showed that visual cues alone could yield the illusion, but with a lower magnitude than that observed in haptic-only or bimodal conditions. Valdez and Amazeen (2008) suggested that vision could augment the basic haptic illusion by virtue of visual facilitation of size perception.

Multi-modal interactions occur for higher-order dimensions as well as their unitary components. We have argued that stiffness perception cannot be directly predicted from force and displacement in a single modality, and the same is true for multi-modal stiffness perception, considered next.

2.2 Visual-Haptic Stiffness Perception

In this section we turn from multi-modal interactions affecting the perception of the components of stiffness, to the higher-order percept itself, which requires combining the components of force and position.

2.2.1 Integration of Visual and Haptic Cues in Stiffness Perception

Multi-modal contributions to stiffness perception were examined by Varadharajan et al. (2008). Using a high-fidelity haptic force-feedback device along with a visual display, they created a simulation of 3D virtual springs that can buckle and tilt in any direction as force is applied. In a stiffness magnitude-estimation experiment, participants freely explored a set of 12 randomly ordered springs with rigidity ranging from 12.0 to 48.0 N/mm. After having interacted with a spring, participants rated perceived stiffness using any number, with the rule that higher numbers meant that the spring felt stiffer. A monotonic relationship between the judged and rendered stiffness was evident. More importantly for present purposes, participants were tested in haptic-only and haptic-visual conditions, and no significant contribution of vision to the perception of stiffness magnitude was found.

In contrast, the same authors found that stiffness discrimination performance was improved by adding visual rendering of the interaction. When vision was excluded in the haptic-only condition, the JND increased by over 20 % relative to the haptic-visual condition, as shown in Fig. 2.2. Thus while vision failed to change the relation between perceived and physical stiffness magnitude, it did provide greater sensitivity in the discrimination of stiffness. Presumably this arose from the kinematics of interaction and changes in size and shape of the spring, although the quantitative weight given to visual versus haptic cues cannot be determined from this study.

The MLE model suggests that the relative weight given to each modality in a multi-modal situation should depend on the relative reliability of the estimates (i.e. inverse variance). In this way, there will be always an advantage for integrated estimates in terms of reliability. This prediction does not appear to be confirmed in a

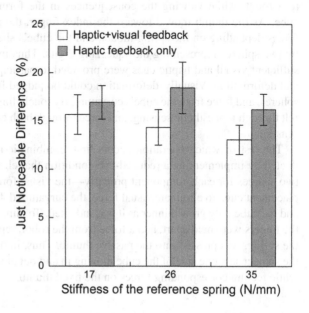

Fig. 2.2 Differential threshold for discriminating spring stiffness was reduced with visual feedback (adapted from Varadharajan et al. 2008)

study by Srinivasan et al. (1996). They investigated the impact of visual information over kinaesthetic information in a comparison of two springs rendered in a virtual environment with a 3 DOF haptic interface. The relationship between visual and haptic stiffness ranged from consistent to reversed (i.e. visual deformation for a hard spring at a particular force level was depicted in conjunction with a haptic rendering of a soft spring at that force level). They found that visual cues essentially dominated kinaesthetic cues. Another effect of vision may be to mitigate biases induced under haptic perception. Along these lines, Wu et al. (1999) found that vision reduced a haptic bias to feel more distant objects as softer.

2.2.2 Models of Visual-Haptic Integration in Stiffness Perception

Higher-order properties require multiple component properties to be *combined*. In the case of stiffness, a ratio is computed between force and displacement. Multi-modal inputs add to the processing burden by imposing the need to *integrate* information from the multiple modalities. This raises an interesting question: When does cue combination occur in relation to integration, or correspondingly, what is integrated/combined? Are individual unidimensional properties (force, displacement) estimated by each modality, integrated across modalities, and then combined? Or does each modality independently combine the components into a higher-order property (stiffness), followed by integration of the multiple estimates of the higher-order that ensue? The models are illustrated in Fig. 2.3.

This question was addressed by Kuschel et al. (2010) using a visual-haptic stiffness display. Participants pinched a virtual spring between two fingers, each attached to a robot, while viewing the consequences in the form of a deformable virtual cube. As the thumb moved toward the index finger, the robots produced resisting forces depending on the spring model, and the cube's shape was visibly deformed by two spheres representing the squeezing digits. Thus in the normal active mode, sufficient visual and haptic cues were provided to participants for estimating force and deformation. Visually, deformation could be judged from the distance between spheres, and force from the cube's curvature. By touch, finger displacement could be felt through kinaesthetic sensing, and force through both cutaneous and kinaesthetic sensing.

The test of which operation occurs first, combination or integration, was conducted by implementing a reduced-cue condition that eliminated the weaker of the two sources for each component property—the visual force cue and the haptic displacement cue. To eliminate visual force, the curvature deformation was eliminated, and the cube only grew thinner as it was indented. To eliminate haptic displacement, the fingers were held apart, and a force from the robot, representing resistance from the spring, was pressed into the passive thumb. Thus, in the reduced-cue condition, the subject saw one side of the cube moving in and out along a sinusoidal trajectory, while feeling corresponding forces on the fixed thumb.

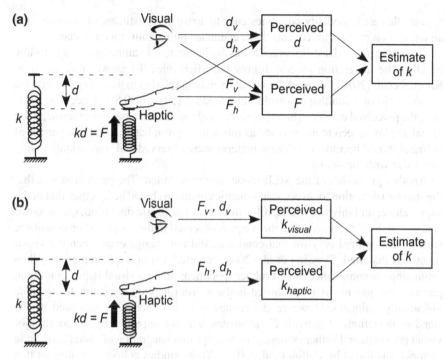

Fig. 2.3 Two processing models have been proposed to explain the visual-haptic integration in stiffness perception (Kuschel et al. 2010). Model (a) suggests that individual properties (i.e., force and deformation) are estimated multimodally and then used in the estimation of stiffness. In contrast, Model (b) suggests cross-modal integration follows the process of cue combination within each modality. That is, stiffness is estimated from unimodal information obtained through vision or touch, followed by a process of cross-modal integration

The reasoning behind such a test is as follows: If combination within each modality into a higher-order property precedes integration, the reduced-cue condition should essentially preclude stiffness from being perceived. That is because both displacement and force are needed to compute the within-modality stiffness estimate; lacking even the weaker cue, the task cannot be done: In the reduced-cue condition, visual cues are available for deformation but not for force, while haptic feedback provides only information of force but no finger displacement. In contrast, if integration precedes combination, the reduced-cue condition should have little negative impact on performance. The weaker cue of each modality can be discarded and the stronger ones (visual displacement and haptic force) can be combined to produce a stiffness estimate. Essentially, there is no integration in this case, because each component property is computed by a single modality, using its stronger cue.

The results of the study supported combination before integration: When the JND was computed from a standard psychophysical task, the value was 0.29 in the active condition and 0.83 in the reduced-cue condition; nearly a 300 % increase. Thus it appears that each modality combines cues to arrive at an estimate of stiffness, and an integration process transforms those estimates into a multi-modal value.

Given that unimodal stiffness estimates are integrated, a natural question to follow is: Does the integration process follow the MLE rule? To answer this question, Kuschel et al. (2010) distorted the visual display in relation to the haptic, so that the relative ratio of visual:haptic displacement was 2:1, 1:1, or 0.5:1. Under the MLE rule, the perceived estimate should move toward the visual value. For example, if the visual display is seen to move twice as much for a given haptic value, the perceived stiffness should become less. These patterns were observed and were quantitatively consistent with the model.

Another prediction of the MLE model however, failed. The prediction was that the standard deviation for every visual-haptic condition should be less than that of the haptic alone, and this was violated for the 2:1 visual:haptic distortion. Quantitative analysis of the data showed that the weights observed in the no-distortion condition, which were optimal only for that condition, did not change even when the visual input was distorted. Drewing et al. (2009) reported a similar phenomenon. When estimating the compliance of soft rubber specimens under a visual-haptic condition, participants' judgments were shifted halfway from haptic-only estimates towards vision-only estimates. However, the reliability of judgments, as measured by the standard deviations of individual's estimates, was not improved by the addition of visual information. Further evidence for non-optimal integration of visual and haptic softness was found by Cellini et al. (2013). These studies collectively suggest that visual and haptic softness cues may not be integrated optimally, particularly when the two inputs are not in congruence.

2.3 Applications of Research on Visual-Haptic Stiffness Perception

A practical problem, particularly in medicine, is how to help people better perceive stiffness through effective augmentation of visual and haptic cues. For example, when performing ultrasound examinations for breast cancer, radiologists compress the target area with the ultrasound probe, observe the ultrasound video, and feel the resisting force to detect possible tumours as changes in tissue stiffness. Research has shown that the human ability to distinguish stiffness is limited. Jones and Hunter (1990) reported an average JND of 23 % for participants haptically comparing the stiffness of simulated springs using a contralateral limb matching procedure. Tan et al. (1995) found a low JND of 8 % for compliance discrimination in a fixed-displacement condition and a significantly higher JND (22 %) when the displacement was varied across the stimuli. For visual discrimination of stiffness, Wu et al. (2012) presented to their participants simulated ultrasound with different levels of speckle noise and structural regularity, and reported JNDs ranging from 12 % to 17 %. In clinical practice, such perceptual limits are reflected in the limited sensitivity of palpation screening. The reported detection rate is only 39–59 % for breast cancer examination (Shen and Zelen 2001).

Engineering platforms have been developed to enhance stiffness perception by the augmentation of visual, haptic, or both types of cues. Based on the results of Kuschel et al. (2010), one approach to augmentation is to facilitate the within-modality perception of stiffness, for example, to augment visual force cues for improved visual perception of stiffness. In an effort toward this goal, Bethea et al. (2004) evaluated a visual force feedback system developed for the da Vinci Surgical System, in which the intensity of the force at the tip of the surgical instrument was indicated using a colour bar varying from green to yellow to red. Their results showed that with such visual force feedback, surgeons could perform robot-assisted surgical knot tying with more consistency, precision, and greater tension to suture materials without breakage. Similarly, Horeman et al. (2012) augmented the display of their laparoscopic training platform with an arrow that continuously informed the trainee about the magnitude and direction of applied force. Their experimental results demonstrated that such a visual representation led to significant improvements in the control of tissue-handling force, not only during the training sessions but also in the post-training tests. Although these studies did not directly assess the perception of stiffness, one would expect that more precise control of force might be associated with enhanced perception of tissue mechanical properties, including stiffness.

Alternatively, haptic cues can be augmented (for an in-depth discussion of haptic augmentation, see Chap. 12). Many tasks require a steady hand, particularly in delicate surgical procedures like microsurgical operations, and therefore the goal of engineering intervention is often to reduce, rather than enhance, the haptic force cues arising from action. The opposite situation also arises, however, when the interaction forces of surgery are imperceptible, and access to force information could be useful to the surgeon. Accordingly, considerable engineering effort has been devoted to developing devices to *augment* force, in order to help the user better perceive it and hence the inherent stiffness of the interaction. One such example is a hand-held force magnification device developed by Stetten et al. (2011), as shown in Fig. 2.4. The device includes a handle held by the user, a sensor and an actuator placed at two ends of the handle, a brace for mounting the device on the user's hand, and a control system. A sensor is mounted on the tip of the handle to measure the pushing/pulling force of the interaction. At the other end, a stack of permanent rare-earth magnets are placed inside the handle and inserted into a custom solenoid. The solenoid is powered by an electrical current from the control box, inducing a Lorentz force on the magnets/handle that is proportional to and in the same direction as the input force. The entire device is mounted to the back of the user's right hand for high portability and manipulation capacities. After careful calibration, the device could amplify the sensed force up to 5.8 times. Jeon and Choi (2009) developed a stiffness-modulation system with a similar goal. Their system could alter the subjective stiffness of soft objects to a specific value by measuring the deformation and then accordingly applying additional forces to the user's hand.

In contrast to the amount of research on design and implementation of augmentation devices, relatively little research has examined the perceptual effectiveness of such augmentation and investigated the factors that may influence the utility of augmented haptic and visual feedback. Consider first the augmentation of

Fig. 2.4 An example of a force augmentation device. The Hand-Held Force Magnifier (Stetten et al. 2011) uses a sensor to measure force F at the tip of the handle, which is amplified to produce $F^* = gF$ in the same direction on the handle using a solenoid mounted on the back of the hand. The total force felt by the user is then $(F + F^*)$ with an amplification of $(1 + g)$. (Adapted from Stetten et al. 2011)

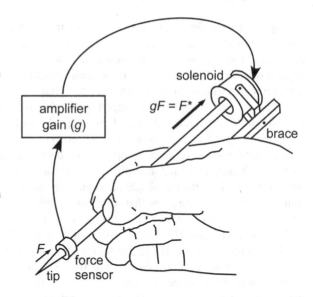

force cues by visual feedback. While it is technically easy to use colour scales or arrows to present force information, the perceptual effectiveness of such feedback is limited, because interpretation requires cognitive mediation. Furthermore, given humans' poor ability to make absolute judgments of colour and length, colour scales or arrows provide the user with little more than a heuristic to estimate the force intensity. An additional consideration is that in order to avoid occlusion of the field of surgery, visual force feedback is often shown at a displaced location, and such spatial displacement could hinder the integration of visual and haptic cues (Gepshtein et al. 2005; Klatzky et al. 2010).

Haptic augmentation of force has the advantage that the augmented signal occurs in the natural perceptual modality, but it is also limited by the sensory and perceptual capacities inherent in the sense of touch. Stetten et al. (2011) characterized their haptic AR system and assessed its effectiveness in judgments of stiffness. Their results showed that the force augmentation induced by the device was well perceived by their participants, leading to significantly higher subjective estimation of stiffness, as shown in Fig. 2.5a. The augmentation effectiveness, however, was found to be greatest for the softest stimuli and to decline gradually with stiffness. The subjective estimates, if related to an up-scaling of the stiffness stimuli by the magnifier's actual power, fell below the control curve and diverged more from it as the stimuli increased (Fig. 2.5b). The change in the augmentation effectiveness may be partly accounted for by a shift in utilization of haptic cues. Small forces are sensed mostly through the deformation of the skin, whereas the perception of large forces relies more on the information from receptors in muscles, tendons, and joints that have relatively lower sensitivity. Jeon and Choi (2009) also conducted an experiment to evaluate the perceptual performance of their stiffness modulation system and found the modulation was felt to be stiffer than the desired value. The authors attributed such error to a lag in haptic rendering. Although these results necessarily reflect the limitations of the devices used, they provide data more generally relevant to the mechanisms underlying stiffness perception.

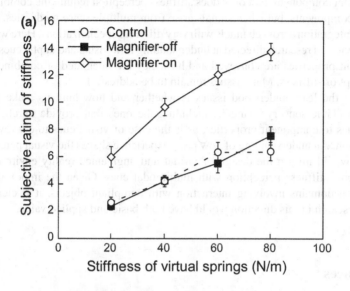

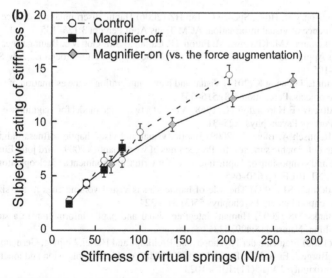

Fig. 2.5 Perceptual effectiveness of the Hand-Held Force Magnifier (Stetten et al. 2011) in stiffness estimation. **a** Mean estimates of perceived stiffness without the device (the control condition) or with the device turned on or off. **b** Re-plot of the magnifier-on data versus the prediction given by the augmented force feedback. The magnifier-on curve fell below the control curve and diverged more from it as the stimuli increased, indicating a gradual declination in effectiveness

2.4 Conclusion

Stiffness, or its inverse compliance, is one of a number of properties that can be called "higher-order", in the sense that it is computed from more than one physically independent component. Not only does stiffness perception require the combination of these components, but when signals arise from multiple sense modalities, it also requires integration across channels with very different specializations. Here we have briefly reviewed research directed at understanding how visual and haptic cues to the component properties are conveyed and how they are integrated or combined into the percept of stiffness. Many issues remain to be addressed.

One of the least understood issues is whether and how humans make use of all the available sensory channels, including the ones that provide unreliable or biased cues to component properties, as is the case of visual cues to convey force information. An understanding of how prior experience shapes the visual perception of force would inform the design of virtual and augmented reality environments that support stiffness perception with multimodal cues. Given the importance of application domains involving interaction with compliant objects, it is clear that further research in this direction would have both basic and applied value.

References

Barbagli F, Salisbury K, Ho C, Spence C, Tan HZ (2006) Haptic discrimination of force direction and the influence of visual information. ACM Trans Appl Percept 3:125–135

Bethea BT, Okamura AM, Kitagawa M, Fitton TP, Cattaneo SM, Gott VL, Baumgartner WA, Yuh DD (2004) Application of haptic feedback to robotic surgery. J Laparoendosc Adv Surg Tech 14(3):191–195

Cellini C, Kaim L, Drewing K (2013) Visual and haptic integration in the estimation of softness of deformable objects. Perception 4(8):516–531

Curtis DW, Attneave F, Harrington TL (1968) A test of a two-stage model for magnitude estimation. Attention Percept Psychophys 3:25–31

Drewing K, Ramisch A, Bayer F (2009) Haptic, visual and visuo-haptic softness judgments for objects with deformable surfaces. In: Proceedings of world haptics 2009, third joint EuroHaptics conference and symposium on haptic interfaces for virtual environment and teleoperator systems. Piscataway, NJ, IEEE, pp 640–645

Ellis RR, Lederman SJ (1993) The role of haptic versus visual volume cues in the size-weight illusion. Attention Percept Psychophys 55(3):315–324

Ernst MO, Banks MS (2002) Humans integrate visual and haptic information in a statistically optimal fashion. Nature 415:429–433

Fechner GT (1860) Elemente der Psychophysik. Breitkopf and Härtel, Leipzig, Germany

Gepshtein S, Burge J, Ernst MO, Banks MS (2005) The combination of vision and touch depends on spatial proximity. J Vis 5(11):1013–1023

Horeman T, Rodrigues SP, van den Dobbelsteen JJ, Jansen FW, Dankelman J (2012) Visual force feedback in laparoscopic training. Surg Endosc 26(1):242–248

Jeon S, Choi S (2009) Haptic Augmented reality: taxonomy and an example of stiffness modulation. Presence Teleoperators Virtual Environ 18:387–408

Jones LA (1986) Perception of force and weight: theory and research. Psychol Bull 100:29–42

Jones LA, Hunter IW (1990) A perceptual analysis of stiffness. Exp Brain Res 79:150–156

Keyson DV (2000) Estimation of virtually perceived length. Presence Teleoperators Virtual Environ 9:394–398

Klatzky RL, Wu B, Stetten G (2010) The disembodied eye: consequences of displacing perception from action. Vis Res 50:2618–2626

Kuschel M, Di Luca M, Buss M, Klatzky RL (2010) Combination and integration in the perception of visual-haptic compliance information. IEEE Trans Haptics 3:234–244

Lanca M, Bryant D (1995) Effect of orientation in haptic reproduction of line length. Percept Mot Skills 80:1291–1298

Michaels CF, De Vries MM (1998) Higher and lower order variables in the visual perception of relative pulling force. J Exp Psychol Hum Percept Perform 24:526–546

Paré M, Carnahan H, Smith AM (2002) Magnitude estimation of tangential force applied to the fingerpad. Exp Brain Res 142:342–348

Seizova-Cajic T (1998) Size perception by vision and kinesthesia. Attention Percep Psychophys 60:705–718

Shen Y, Zelen M (2001) Screening sensitivity and sojourn time from breast cancer early detection clinical trials: mammograms and physical examinations. J Clin Oncol 19:3490–3499

Srinivasan MA, Beauregard GL, Brock DO (1996) The impact of visual information on haptic perception of stiffness in virtual environments. ASME Dyn Syst Control Div 58:555–559

Stanley G (1966) Haptic and kinesthetic estimates of length. Psychon Sci 5:377–378

Stetten Gl, Wu B, Klatzky R, Galeotti J, Siegel M, Lee R, f Mah F, Eller A, Schuman J, Hollis R (2011) Hand-held force magnifier for surgical instruments. Information processing in computer-assisted interventions. Lecture notes in computer science, 6689, Springer, Berlin. pp 90–100

Stevens JC, Mack JD (1959) Scales of apparent force. J Exp Psychol 58:405–413

Stevens SS (1975) Psychophysics: introduction to its perceptual, neural, and social prospects. Wiley, New York

Sun Y, Hollerbach JM, Mascaro SA (2008) Predicting fingertip forces by imaging coloration changes in the fingernail and surrounding skin. IEEE Trans Biome Eng 55:2363–2371

Tan HZ, Durlach NI, Beauregard GL, Srinivasan MA (1995) Manual discrimination of compliance using active pinch grasp: the roles of force and work cues. Attention Percep Psychophys 57:495–510

Teghtsoonian M, Teghtsoonian R (1965) Seen and felt length. Psychon Sci 3:465–466

Teghtsoonian M, Teghtsoonian R (1970) Two varieties of perceived length. Attention Percept Psychophys 8:389–392

Valdez AB, Amazeen EL (2008) Sensory and perceptual interactions in weight perception. Attention Percept Psychophys 70:647–657

van Beers RJ, Sittig AC (1999) Integration of proprioceptive and visual position information: an experimentally supported model. J Neurophysiol 81:1355–1364

Varadharajan V, Klatzky R, Unger B, Swendsen R, Hollis R (2008) Haptic rendering and psychophysical evaluation of a virtual three-dimensional helical spring. In: Proceedings of the 16th symposium on haptic interfaces for virtual environments and teleoperator systems. IEEE, Piscataway, NJ, pp 57–64

Wang X, Ananthasuresh GK, Ostrowski JP (2001) Vision-based sensing of forces in elastic objects. Sens Actuators A Phys 94:142–156

White PA (2012) The experience of force: the role of haptic experience of forces in visual perception of object motion and interactions, mental simulation, and motion-related judgments. Psychol Bull 138:589–615

Woodworth R, Schlossberg H (1960) Experimental psychology, Revised edn. Henry Holt, New York

Wu B, Klatzky RL, Shelton D, Stetten G (2008) Mental concatenation of perceptually and cognitively specified depth to represent locations in near space. Exp Brain Res 184:295–305

Wu B, Klatzky RL, Hollis R, Stetten G (2012) Visual perception of viscoelasticity in virtual materials. Presented at the 53rd annual meeting of the psychonomic society, Minneapolis, Minnesota, Nov 2012

Wu W, Basdogan C, Srinivasan MA (1999) Visual, haptic, and bimodal perception of size and stiffness in virtual environments. ASME Dyn Syst Control Div 67:19–26

Chapter 3
Vibrotactile Sensation and Softness Perception

Yon Visell and Shogo Okamoto

3.1 Introduction

This chapter describes how mechanical vibrations can affect the perception of several material and surface properties, with an emphasis on the perception of compliance. Vibrations are fluctuations of force or displacement. They are generated by interactions with objects and as such they are produced during numerous human activities. They accompany, for example, frictional sliding of surfaces, tapping, rolling movements, displacement and compression of granular and aggregate materials, fracturing and breakage processes.

Prior research has demonstrated that the haptic channel is sensitive to vibrations, known as the *vibrotactile* sense, and can be used to discriminate touched surfaces of objects to extract properties such as roughness or surface regularity, and to identify events, such as contact onset and contact slip. Despite its importance, the vibrotactile sense has received little attention to date as a potential cue for compliance perception, especially when compared to other haptic perceptual channels or cues, such as cutaneous contact area, proprioception, and kinesthesia. One reason can be traced to the contact mechanical origin of vibromechanical signals which consist of high-frequency fluctuations in force or displacement. In many of the mechanical interactions listed above, vibration energy is produced through impacts between the surfaces of objects at a macroscopic scale or through interaction between surface microgeometry (asperities) as it happens during sliding friction (Akay 2002). In such cases, materials with high stiffness yield wide frequency bandwidth transient or sustained signal elements. In contrast, for compliant objects the effective stiffness

Y. Visell (✉)
Drexel University, Philadelphia, PA, USA
e-mail: yonvisell@gmail.com

S. Okamoto
Nagoya University, Nagoya, Japan

© Springer-Verlag London 2014

M. Di Luca (ed.), *Multisensory Softness*, Springer Series on Touch and Haptic Systems,
DOI 10.1007/978-1-4471-6533-0_3

of the impacting structures is low and the material may also be more damped. These properties yields only low energy at high frequencies during interaction.

Nevertheless, several studies have suggested that vibromechanical signals can influence perception during object palpation for a wide range of object compliances (Giordano et al. 2012; Kobayashi et al. 2008; Ben Porquis et al. 2011; Takahiro et al. 2010; Okamoto 2010; Ikeda et al. 2013; Rust et al. 1994; Okamura et al. 2001; Kuchenbecker et al. 2006; Kildal 2010, 2012; Visell et al. 2011). It is well established that high-frequency mechanical vibrations generated during manually tapping, scraping with a probe, or scanning with a finger can influence the perception of properties such as hardness (as reviewed below) and roughness (Klatzky and Lederman 1999; Hollins and Risner 2000; Bensmaia and Hollins 2003, 2005; Klatzky et al. 2003; Okamura et al. 1998). For example, amplifying vibrations generated during manual surface scanning, or imposing sinusoidal vibrations, increases perceived surface roughness (Hollins et al. 2000). Thus, it is reasonable to ask whether there exist high-frequency cues that are capable of influencing compliance judgements.

3.1.1 Vibrotactile Sensory Information

Vibromechanical stimulation of the skin affects both cutaneous receptors and receptors embedded in deep tissues, including muscles and tendons (Ribot-Ciscar et al. 1989; Freeman and Johnson 1982; Johnson 2001; Vedel and Roll 1982). The former stimulation include fast-adapting (FA) mechanoreceptors sensitive to phasic signals, in the form of Meissner and Pacinian corpuscles. These mechanoreceptive channels respond to transient or high-frequency mechanical stimuli. Also present are slower adapting (SA) mechanoreceptors that respond primarily to tonic signals produced by sustained or slowly-varying mechanical stimuli. In previous studies, the vibrotactile sense has been particularly associated with FA receptors. Mechanoreceptive afferents, which communicate the neural result of mechanical stimuli to the central nervous system, have been associated with receptive fields near to receptors that they terminate on. The size of the respective receptive fields for FA or SA mechanoreceptive afferents can range from a few square millimeters to several square centimeters, depending on the receptor type, innervation density, and biomechanical factors. Among physiologically identified FA receptors, Meissner corpuscles have small receptive fields, while Pacinian corpuscles lie deeper in the skin and possess larger receptive fields. The skin is sensitive to vibrotactile stimuli over a broad range of frequencies, up to nearly 1,000 Hz. Meissner corpuscles respond preferentially to vibrotactile signals in the range from 10 to 100 Hz, while Pacinian corpuscles respond to higher frequencies, but neither type exhibits narrow frequency-selective tuning like that present in the auditory system. In glabrous skin, which is found on the volar surface of the hand and feet, FA afferents comprise about 70 % of the cutaneous population. Vibrotactile sensitivity, measured in terms of the absolute or difference threshold for detection, varies as a function of body location and stimulus properties including contact conditions and frequency. Surveys of tactile sensitivity at different body locations and for different stimulus parameters, including frequency and amplitude of stimulation, can be found in the following reviews (Morioka et al. 2008; Morioka and Griffin 2002; Verrillo 1966).

3.2 Contact Mechanics and Softness Cues

Haptic compliance perception involves discerning the deformability of objects touched with the hand or foot, or even objects felt using a tool. As discussed in Chap. 1, compliance $C = 1/k$ can be quantified in terms of mechanical stiffness k, which in turn depends on the Young's modulus and geometry of the material. In the simplest case, the deformation x of a material can be described via a linear, quasi-static relation between force $F = -kx$ and displacement x, or between continuum mechanical quantities of stress $\sigma = -E\varepsilon$ and strain ε.

The problem of softness perception consists of using haptic sensations to percep-tually recover the compliance or material elasticity of an object. Thus, the notion of softness involves the extraction of object properties from stresses and strains, or forces and displacements, that are felt during exploration. Most prior research has investigated compliance perception via manual touch (Harper and Stevens 1964; Scott-Blair and Coppen 1940; Freyberger and Färber 2006; Tan et al. 1995; Srini-vasan and LaMotte 1995; LaMotte 2000; Friedman et al. 2008; Bergmann Tiest and Kappers 2009). However, the haptic perceptual system is able to discriminate objects of different compliance in a multitude of ways, including touching with a tool (LaMotte 2000) or with the hand or foot (Giordano et al. 2012; Kobayashi et al. 2008).

The perceptual system is capable of judging softness in different ways depending on the information that is available. The contributions from different sources depend upon the actions, tasks, or exploratory procedures being performed, the properties of the objects (their material composition, geometry, and microgeometry), and the contact mechanical setting involved. Their integration thus must account for such sources of variability and should proceed accordingly (see Chap. 5 for a model). The same can be said about what kind of vibrations are available during different types of interactions and thus how vibrotactual information could be used for softness perception.

It is useful to consider four basic interaction patterns, which are represented in Fig. 3.1. They consist of *direct skin contact* with a compressed elastic object, *indirect skin contact* with such an object, *transient contact* with a touched object, and *frictional sliding*. Through these, it is possible to gain some insight into potential roles of vibrotaction in softness perception. In the next four sections we will analyse the vibrotactile information available in each of these interactions and what experimental results are available about the perception of softness.

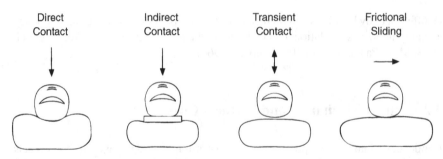

Fig. 3.1 Four prototypical scenarios where vibrotactile information is available and can potentially specify object softness: **a** Compressing a soft object with direct finger contact; **b** Compressing such an object via a rigid intermediary; **c** Producing a short, ransient contact through tapping; **d** Sliding the finger across the object surface

3.2.1 Direct Skin Contact

The first interaction type we consider involves direct contact between the skin and an elastic, compressed object, as shown in Figure 3.1a. This setting can be modelled using the Hertzian theory of non-adhesive linear elastic contact between two compressible bodies (Johnson 1995). One body consists of the skin and pulp of the finger, and the other constitutes the touched object. As the total normal force applied by the finger increases, the area of contact between the bodies grows. Simultaneously, the skin and underlying tissues deform, as can be quantified by an increase in strain energy density near the contact region. This gives rise to at least two potential perceptual cues (Bicchi et al. 2000; Scilingo et al. 2010) (see Chap. 11):

1. The rate of increase of contact area between finger and object surface with the normal force between finger and object surface
2. The rate of increase in strain energy in the volume of the finger near the contact region with normal force

The rate of increase in normal force can often be assumed to be slow, and the accompanying dynamics to be damped, due to the highly viscoelastic nature of the materials involved. Often the mechanics can be modelled as quasi-static. In such cases, transient mechanical signals can be presumed to be insignificant. Commensurate with this assumption, it could be hypothesized that, at a physiological level, the neural input from slowly-adapting (SA) afferents from the finger provide most sensory information, while inputs from fast-adapting (FA) afferent channels are less important. One could argue that vibromechanical cues contribute to softness perception by simulating transient strain patterns over the skin like those produced when touching an object, but there is little evidence to suggest that such cues contribute significantly to softness perception during direct skin contact, perhaps because contact area itself is so highly weighted, when available: As noted in Chaps. 1, 2, and 5, physiological and psychophysical studies have shown that haptic perceptual sensitivity to softness is highest when there is direct skin contact with a deformable surface.

An alternative possibility, discussed further in Sect. 3.3, is that vibration feedback could bias estimates of applied force during object compression.

3.2.2 Indirect Skin Contact

When touch is mediated by a rigid link, such as a handheld stylus or rigid mechanism, the cutaneous perceptual cues (1,2) noted above do not provide information about compliance, since contact area and skin strain reflect the force between the finger and a rigid surface but do not independently evidence the displacement of the surface. Adding a rigid link to the interaction with an object having deformable surfaces makes the interaction equivalent to the one obtained with an object with rigid surfaces. In such cases where cutaneous information is not directly informative about compliance, discrimination performance is significantly reduced (Srinivasan and LaMotte 1995). With rigid surfaces compliance estimation requires the combination of force and displacement information (see Chaps. 1 and 5). To estimate object compliance during indirect touch, cutaneous force cues could be combined with displacement information obtained from visual and proprioceptive sense data (see Chaps. 1, 2, and 5 for more information).

Several studies have demonstrated that individuals are able to estimate compliance under such settings (for example, estimating the compliance of a spring-loaded mechanism) (Srinivasan and LaMotte 1995; Tan et al. 1995; Bergmann Tiest and Kappers 2009; Jones and Hunter 1990). The results generally demonstrate reduced sensitivity when compared to the case of direct skin contact with a deformable object is available.

What does this suggest about possible roles of vibrotactile sensation in compliance estimation via indirect touch? It could be hypothesized that in such a setting, vibration could affect compliance estimates in one of two ways: by either biasing estimated displacements Δx or estimated forces ΔF (Fig. 3.2). As noted in Chap. 5, both types of bias are possible.

As vibrations are normally produced during object deformation, amplifying these vibrations could lead to a change in perceived compliance. This is partly supported in the literature (Visell et al. 2011; Kildal 2012, 2010).

Fig. 3.2 Vibration stimuli may bias compliance estimates by altering perceived force F or displacement x

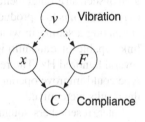

3.2.3 Transient Contact

So far we have discussed potential increases in perceived compliance due to vibration. In principle, however, vibrotactile cues could have the opposite effect namely they could decrease perceived compliance. Tapping on a hard surface elicits characteristic vibrations, in the form of transient mechanical signals, with broad frequency content, due to the rapid changes in contact forces. During such an interaction, a stiffer object may yield a more perceptually prominent vibrotactile signal. A model of the forces involved can be given by the Hertz theory of viscoelastic contact. A simplified version of the Hertz model that is suitable for the analysis of transient contact forces during impact with a viscoelastic object is due to Hunt and Crossley (1975), and can be written as

$$f_{impact} = K(z) - D(z)\dot{z}, \tag{3.1}$$

where z represents the depth of penetration beyond the undeformed surface of the object, K models the growth rate of contact surfaces, and D captures the dissipation. The tapping force excites vibrations in the object that can often be described by a source-filter model, consisting of an impulse response $h(t)$ equal to the response to an impulsive force, so that the net displacement is given by $y(t) = z(t) * h(t)$, where $*$ denotes convolution in time. For a stiff object this response combines contributions of broadly distributed frequency content arising from the contact force and contributions of high-frequency resonant modal frequencies. Either source may lie within the range of frequencies humans are sensible to, hence they could provide a potential perceptual cue to contact and to object compliance. Indeed, the notion that tapping on a surface is a suitable action for exploring surface hardness is familiar from everyday experience, and has further been explored experimentally (Lederman and Klatzky 1987).

The transient forces that are generated during tapping can yield vibromechanical signals that can be readily reproduced via a haptic interface using sufficiently wide bandwidth motors or actuators. Among the earliest work exploring the use of contact-generated vibration cues to communicate information about touched objects are robotic teleoperation studies in which a human operator of a master robot uses a slave robot to manipulate objects in a remote environment. The operator is provided with vibrotactile feedback that reproduces accelerations measured near the end effector of the slave device (Massimino and Sheridan 1993; Kontarinis and Howe 1995). The goal of such an arrangement, which can be described as a form of sensory substitution (Visell 2009), is to reproduce transient accelerations experienced at the end effector, simulating a setting in which the master and slave device were coupled via a rigid link capable of transmitting high-frequency vibrations, in the frequency range of several hundred Hz. The research investigated the extent to which feedback of this type could improve operator performance on basic tasks, such as peg insertion, but the results were mixed.

Later, researchers sought to enhance force or contact information in computer simulated virtual environments with vibrations that were designed to mimic the physical response of real objects, either by means of a physical model (for example, the Hunt-Crossley impact model mentioned above) or based on the measured response of tapped materials (Kontarinis and Howe 1995; Okamura et al. 2001; Kuchenbecker et al. 2006). Studies have shown that by superimposing transient vibrations on contact forces, perceived surface hardness can be increased and material identity can be modified or enhanced (Kuchenbecker et al. 2006). Similar results have been observed with transient audio feedback, in the form of tapping sounds (see Chap. 4 for a comprehensive review). In general these methods of rendering, which employ a signal delivered through the auditory, visual, or tactile sense modalities which are triggered by the movement of the participant, have collectively been given the name of "event-based haptics".

3.2.4 Frictional Sliding

Friction involves tangential forces produced during the sliding of objects. Texture refers to small-scale modifications of mechanical interaction responses during sliding or during indentation. The forces involved comprise both slowly-varying nominal or constitutive responses and fluctuating components generated by surface or material imperfections. The latter signal components relate relative displacement of the objects concerned to high-frequency frictional force components (Ibrahim 1994; Akay 2002), whose frequency bandwidth can overlap that of the vibrotactile sense. In principle, perceptual information about interaction parameters, such as applied force or displacement, and material properties, such as surface hardness, are available through such signals. In everyday terms, even for objects with very soft surfaces, such as textiles like velvet or silk, texture-like force components can provide information about material properties. Additionally, as demonstrated in the well-known parchment skin illusion (Jousmäki and Hari 1998), amplifying the sound of frictional rubbing can create a perceptual experience that the sliding surface is dryer, rougher, or harder than is nominally the case.

In contrast to softness sensations elicited by pressing on a surface, softness cues produced by stroking with the finger are more difficult to interpret, since it is more challenging to analyze the physical interactions between the finger (or a probe) and the surface, and thus to relate softness perception to surface parameters. Nonetheless, it is plausible that individuals may use information acquired by stroking or scanning with a finger in order to estimate object compliance. First, because the sliding dynamics may directly depend on bulk material properties such as elasticity. Second, because surface properties may elicit prior expectations for object softness.

Textile softness is often perceived via rubbing with the fingers, as has been extensively studied in areas of the literature on applied perception and ergonomics (Pense-Lheritier et al. 2006; Chen et al. 2009). In order to obtain objective measures of fabric softness, several researchers have investigated the relation between reported textile softness and vibromechanical cues. Rust et al. (1994) were able to predict textile softness ratings using vibromechanical measurements obtained from a novel engineering instrument. Lang and Andrews (2011) observed a connection between the object rigidity and sliding-produced vibrations in a probe . These studies were motivated by the idea that stroking a harder material can lead to a larger microscopic movements of a rigid probe as it comes in contact with the microscale defects of the surface upon which it slides. The magnitude and frequency of the vibrations caused by these contacts depend on the surface hardness—i.e., the microscale defects of the surface of a compliant materials would deform rather than making the probe. When the probe is mechanically coupled with a hard surface, their collective rigidity relatively increases, which leads to a larger resonance frequency and vibratory accelerations. When the probe is coupled with a soft surface, their comprehensive rigidity decreases, which leads to a smaller resonance frequency and significant damping ratio. Hence, the mean acceleration values depending on contact forces approximately reflect the compliance of surfaces. What is still unclear is whether humans can judge softness based on the cues generated by stroking alone as it is the case instead of perceived softness by tapping.

3.3 Effects of Low-Frequency Vibration on Softness

Prior literature has indicated that low-frequency vibrotactile stimuli, in the frequency range from 3 to 5 Hz, can evoke the perceptual sense of material softness (Ben Porquis et al. 2011; Takahiro et al. 2010). The softness experience that is evoked by such stimuli grows slowly, with a percept of vaguely defined onset. This slow but still noticeable softness sensation could be of practical interest, because it holds the potential to provide any handheld or grounded devices having vibrotactile channels with added value, namely the distinctive ability to produce artificial softness sensations.

3.3.1 Prospective Mechanism

Skin-mediated softness percepts are produced by the spatial distribution of pressure on the skin. Intensive pressure on a small area results in an experience of the contact with hard object, whereas widely extensive pressure is perceived as that with soft material, as shown in Fig. 3.3. According to the Hertzian contact theory (Johnson 1995), the radius a of a circular contact area made by two spherical bodies, here representing the finger pad and a soft object, is

$$a = \left(\frac{3w}{4}\frac{R}{E}\right)^{1/3}, \tag{3.2}$$

where w, R, and E are the contact force, composite radius and Young's modulus of the two bodies. The composite variables are

$$\frac{1}{R} = \frac{1}{R_f} + \frac{1}{R_s} \qquad (3.3)$$

and

$$\frac{1}{E} = \frac{1 - v_f^2}{E_f} + \frac{1 - v_s^2}{E_s}, \qquad (3.4)$$

where R_i, E_i, and v_i are the radius, Young's modulus, and Poisson ratio of body i. The suffixes f and s describe the finger and surface to be touched by the finger, respectively. The values of R_f, E_f, and v_f are assumed to be known. This can be justified by the assumption that a human observer should be roughly familiar with their own finger pad's size and softness. In case the soft object is a flat surface, $R_s = \infty$ and $R \approx R_1$. In this case, the two softness parameters v_s and E_s may be estimated from a and w. Note that the value range of Poisson ratio is narrow and typically near 0.3. The effect of v_s^2 can therefore be viewed as insignificant compared to E_s. The cutaneous and kinesthetic receptors of the human finger are capable of estimating both contact area and applied force. Humans can make use of slowly adapting type I (SA I) mechanoreceptive units and pressure-sensitive nocireceptors distributed beneath skin to estimate the pressure distribution and applied force caused by contact with soft surfaces. Receptors in the muscles and tendons of the finger and wrist are also sensitive to forces applied to finger pad.

At least three different engineering interfaces have been designed around the aforementioned principles, albeit by means of very different devices (see Chap. 14). Bicchi et al. (2000) and Scilingo et al. (2010) fabricated a finger pad contactor consisting of several concentric actuated cylinders with different radii. This device made it possible to control the pressure and contact area between a finger pad for testing the hypothesis that contact area plays a key role in softness percepts. In contrast, Fujita and Ohmori (2001) and Kimura et al. (2010) used balloon and sheet-based tactile displays, respectively, to elucidate and demonstrate the effects of contact area on softness percepts. The results of these studies affirmed the primary contribution of contact area to softness perception, although the explanatory hypotheses that each proposed have not been completely unanimous. Nonetheless, changes in the contact area can be said to effectively influence softness perception.

It is evident that the pressure distribution in the contact area has a strong connection with softness perception whereas its specific role leaves room for discussion. As described above, SA I units and some nociceptors are sensitive to pressure or sustained indentation. There is a distinct possibility that the activation of SA I units by vibrotactile stimuli induces softness percepts. For the low-frequency band or static mechanical stimuli, SA I units have the lowest thresholds, i.e., they are more sensitive than other units.

As described in the following section, larger low-frequency vibratory amplitudes are associated with softer percepts. The changes in the indentation can result in changes in deformed skin area, as shown in Fig. 3.3. With large skin indentation, large populations of SA I units are expected to be activated. Additionally, the size of

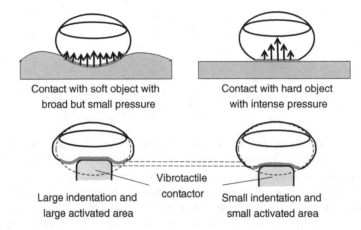

Fig. 3.3 *Top* Pressure distribution caused by contacts with objects. *Down* Finger pad deformation via vibrotactile indentation

population of activated SA I units is more predictive than the impulse rate of single unit about the area of the contact (Suzuki et al. 1999).

3.3.2 Low-Frequency Softness Rendering

Fig. 3.4 shows a schematic view of an experiment that evaluated effects of low-frequency on softness perception (Okamoto 2010; Takahiro et al. 2010). The experiment was based on the method of constant stimuli. In it, participants compared a low-frequency (5 Hz) vibrotactile stimulus and a physically soft sample and judged which stimulus was felt softer. Vibrations were generated using a voice coil

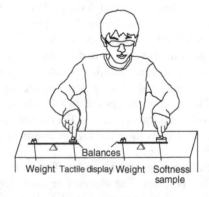

Fig. 3.4 Comparison between low-frequency vibrotactile stimuli and softness specimen

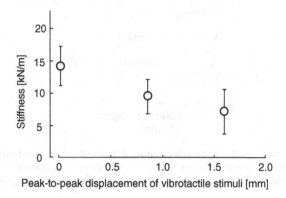

Fig. 3.5 Low-frequency vibrotactile stimuli versus softness of specimen

motor with a contactor consisting of a plastic plate. The vibrotactile amplitudes were
0–1.6 mm under a sustained load of 0.5 N. The physically soft specimens were cylindrical silicone rubber samples whose spring coefficient ranged 4.7–22.3 N/mm. Two
balances were used to ensure equal loads were applied to the two stimuli. Fig. 3.5
shows the compliance of the silicone perceived to be matching the one of vibrating
device. The perceived hardness decreased as a function of the vibration amplitude. In
this experiment, the true contact area between the vibrotactile contactor and finger pad
varied with stimulus amplitude in such a way larger amplitudes led to larger contact
areas. In a subsequent investigation, the contact area was held constant, and the same
effects of the amplitudes on the softness percepts were observed (Ben Porquis et al.
2011). Furthermore, there is evidence that low-pass-filtered white-noise vibrations
can have a similar impact on softness percepts (Ikeda et al. 2013).

3.4 Volumetric Softness

During direct and indirect object contact (Fig. 3.1), vibration can accompany the
compression when the displacement of the volume releases energy. Such vibration
can be a potential cue to softness, but vibration can be released during a variety
of other inelastic processes. It is thus not surprising that vibromechanical energy
alone, when is not correlated with action, is unlikely to affect perceived softness. For
example, when touching a washing machine, one is seldom left with the impression
of owning a "soft" appliance.

A haptic interface that provides similar vibration feedback to what is felt during such situations might be expected to yield increased sensations of softness. A
schematic illustration is shown in Fig. 3.6. Several published demonstrations and
perceptual studies have provided evidence that vibrotactile feedback can modulate volumetric softness (Visell et al. 2011; Kildal 2010, 2012). Other perceptual

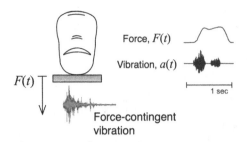

Fig. 3.6 The compression of a surface accompanied by suitable vibration feedback can yield increased sensations of volumetric softness, due to a straightforward sensorimotor contingency between the generation of internal vibromechanical energy during object compression

effects, including force-to-visual displacement gain modulation [or "pseudo-haptics" (Lecuyer et al. 2000)] are also known to affect volumetric softness perception.

In one study of effects of vibration feedback on volumetric compliance perception, Kildal demonstrated that a rigid box pressed with the finger or a stylus could feel as though it compressed in height (Kildal 2010, 2012). This sensation was evoked by vibration feedback that was coupled to applied force. The stimuli consisted of vibrations supplied by a resonant vibrotactile actuator driven by a voltage (amplitude) signal. The stimulus design was motivated by a mechanical model consisting of a spring loaded mechanism moving over a corrigated surface. A vibration transient was supplied whenever the normal force on the device changed by a quantity ΔF. Averaging over vibration transients yields a stimulus $s(t)$ with RMS amplitude given by

$$s_{\mathrm{rms}}(t) = S_0 \frac{dF}{dt} = S_0 k \frac{dx}{dt}, \tag{3.5}$$

where S_0 is a constant amplitude factor, F is normal force, t is time. The parameters k and x are the virtual stiffness and displacement of the simulated mechanical model responsible for producing the feedback.

Visell et al. (2011) investigated whether action-synchronized vibromechanical stimuli felt when pressing on a surface could yield influence the perceived compliance of a walking surface. The vibration feedback they presented to participants depended on the force that was applied to the walking surface. These signals were comparable to those experienced when stepping on a natural material that produces acoustic energy when compressed (e.g., snow, gravel, leaves, soil), or displacing a foot operated mechanism that generates friction noise (e.g., a pedal or slider). Vibromechanical results of stepping onto a walking surface have been found to be related for the identification of the type of material that natural and man-made walking surface are composed of (Giordano et al. 2012). The authors' investigation was based on a pair of experiments. The first sought to ascertain the dependence of perceived compliance on on the vibration stimulus waveform, and on the relation between applied force and feedback amplitude. The stimuli consisted of vibration feedback that varied in two respects: the driving waveform (sinusoidal, white noise, or poisson noise) or

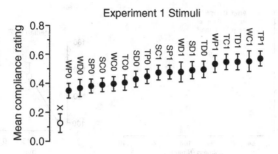

Fig. 3.7 In one experment, Visell et al. (2011) found that nine different types of vibration feedback were all able to elicit a significantly increased percept of compliance when compared with the "no vibration" case. Image from Visell et al. (2011)

and the temporal dependence on force (proportional, time-derivative, or constant). Results indicate that vibrotactile feedback could elicit an increase in perceived compliance and that the effect grew in proportion to the feedback amplitude (Fig. 3.7). In a second experiment, the authors manipulated the compliance of the walking surface that users stepped on, using a novel haptic interface (Fig. 3.8). The goal was to determine whether compliance and vibration are perceptually integrated in an organized way, and to calibrate the perceptual effect in physical units, yielding a quantity of change in physical compliance that could be achieved with a given vibrotactile feedback amplitude. Results point out that vibrotactile feedback provides a perceptual cue for compliance (Fig. 3.9). The vibration amplitudes required to produce an appreciable bias in perceived compliance were very low, near to the absolute threshold at which subjects could detect the vibrotactile stimuli when stepping. Thus, the

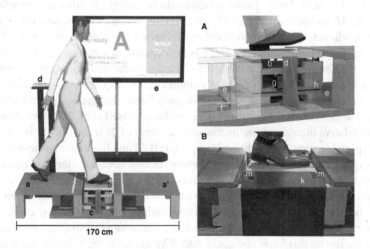

Fig. 3.8 Effects of vibration on volumetric softness. Image from Visell et al. (2011)

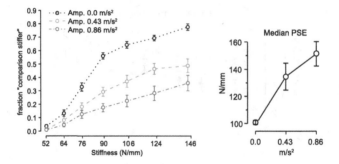

Fig. 3.9 Effects of vibration on volumetric softness

amplitudes used were considerably smaller than the ones experienced during normal walking on natural granular materials, such as gravel, as humans are well aware of the vibrations present there. It is not entirely clear why the stimuli were so effective in this experiment, which did not involve training and did not require awareness that vibration feedback was being provided. The compliance estimation task adopted in this study resembles prior experiments in which subjects used their hands to estimate the haptic compliance of spring mechanisms or other objects with non-deformable surfaces (LaMotte 2000; Tiest and Kappers 2009; Jones and Hunter 1990). Based on those results, and on considerations of contact mechanics, it was expected that subjects in the experiments described here required both force and displacement information in order to judge compliance. In this light, it appears that added vibration feedback results in a modification of force and/or displacement information that increases compliance estimates. Prior research has shown that localized vibration stimulation of the foot sole can increase perceived force at the same location (Kavounoudias et al. 1999, 1998), which could be thought to influence compliance judgments. However, an increase of force estimates would tend to reduce compliance estimates, whereas the aforementioned results reviewed show an opposite tendency. Thus, a more likely explanation, which is also consistent with the results of Kildal (2010, 2012), seems to be that vibration elicited an increased sensation of displacement, as if an object or material was compressed or displaced. If so, an observer could be presumed to infer an increase in compliance that grows linearly with the increased sensation of displacement. The results of (Visell et al. 2011) were consistent with relative increases in estimated displacement of 25.0 and 33.5 % in low- and high-amplitude vibration conditions (amplitude 0.43 and 0.86 m/s^2) as tested in the study, suggesting a monotonically increasing relation between vibration amplitude and compliance perception. The model proposed in Chap. 5 suggest that the increase in sensed displacement may be related to a biased detection of the unperturbed position of the object. The first contact with a vibrating object is obtained earlier than the resting state position. This should increase the amount of displacement sensed afterwards. But on this basis one could ask: Why indeed is the washing machine not perceived to be "soft"?

3.5 Conclusions

A wide variety of mechanical signals arise during haptic interaction with structured surfaces and objects. Among these, there are several sources of high-frequency mechanical vibration that might influence haptic softness perception. They range from transient vibrations induced during tapping on a surface, to subtle fluctuating forces during frictional sliding, to vibrations resulting from inelastic processes accompanying the compression of solid objects. The heterogeneous nature of the contact mechanical interactions involved precludes, to some extent, a unified explanation of all such effects. However, several studies reviewed above have provided evidence for such effects. Together, they indicate that vibrotactile cues can influence the perception of object compliance. This is notable because in standard accounts of haptic perception, such cues are not normally considered to be relevant to softness perception. Rather, compliance is often described as being primarily mediated via proprioceptors in the muscles and joints, and via cutaneous receptors for force or contact deformation.

In most cases, vibrotactile cues have a comparatively weaker influence on perceived softness than can be achieved by manipulating the material properties of the objects themselves or by manipulating force-displacement relations reproduced by a haptic interface. Nonetheless, the notion that a hard surface can be rendered more or less compliant through mechanical vibrations alone is powerful and compelling.

References

Akay A (2002) Acoustics of friction. J Acoust Soc Am 111:1525

Ben Porquis L., Konyo M, Tadokoro S (2011) Representation of softness sensation using vibrotactile stimuli under amplitude control. In: Robotics and automation (ICRA), 2011 IEEE international conference on, IEEE, pp 1380–1385

Bensmaia S, Hollins M (2003) The vibrations of texture. Somatosens Mot Res 20(1):33–43

Bensmaia S, Hollins M (2005) Pacinian representations of fine surface texture. Percept psychophys 67(5):842

Bergmann Tiest WM, Kappers AML (2009) Cues for haptic perception of compliance. IEEE Trans Haptics 2(4):189–199

Bicchi A, Schilingo EP, De Rossi D (2000) Haptic discrimination of softness in teleoperation: the role of the contact area spread rate. IEEE Trans Rob Autom 16(5):496–504

Chen X, Shao F, Barnes C, Childs T, Henson B (2009) Exploring relationships between touch perception and surface physical properties. Int J Des 3(2):67–76

Freeman AW, Johnson KO (1982) Cutaneous mechanoreceptors in macaque monkey: temporal discharge patterns evoked by vibration, and a receptor model. J physiol 323(1):21–41

Freyberger F, Färber B (2006) Compliance discrimination of deformable objects by squeezing with one and two fingers. In: Proceedings of euroHaptics 2006, pp 271–276

Friedman R, Hester K, Green B, LaMotte R (2008) Magnitude estimation of softness. Exp Brain Res 191(2):133–142

Fujita K, Ohmori H (2001) A new softness display interface by dynamic fingertip contact area control. In: 5th world multiconference on systemics cybernetics and informatics, pp 78–82

Giordano BL, Visell Y, Yao H-Y, Hayward V, Cooperstock JR, McAdams S (2012) Identification of walked-upon materials in auditory, kinesthetic, haptic, and audio-haptic conditions. J Acoust Soc Am 131:4002

Harper R, Stevens SS (1964) Subjective hardness of compliant materials. Q J Exp Psychol 16: 204–215

Hollins M, Fox A, Bishop C (2000) Imposed vibration influences perceived tactile smoothness. Perception 29(12):1455–1466

Hollins M, Risner S (2000) Evidence for the duplex theory of tactile texture perception. Percept Psychophys 62(4):695–705

Hunt K, Crossley F (1975) Coefficient of restitution interpreted as damping in vibroimpact. J Appl Mech 42:440

Ibrahim R (1994) Friction-induced vibration, chatter, squeal, and chaos-part i: Mechanics of contact and friction. Appl Mech Rev 47(7):209–226

Ikeda A, Suzuki T, Takamatsu J, Ogasawara T (2013) Producing method of softness sensation by device vibration. In: 2013 IEEE international conference on systems, man, and cybernetics, pp 3384–3389

Johnson KL (1995) Contact mechanics. Cambridge University, Cambridge

Johnson KO (2001) The roles and functions of cutaneous mechanoreceptors. Curr Opin Neurobiol 11(4):455–461

Jones LA, Hunter IW (1990) A perceptual analysis of stiffness. Exp Brain Res 79:150–156

Jousmäki V, Hari R (1998) Parchment-skin illusion: sound-biased touch. Curr Biol 8(6):R190–R191

Kavounoudias A, Roll R, Roll J-P (1998) The plantar sole is a'dynamometric map'for human balance control. Neuroreport 9(14):3247–3252

Kavounoudias A, Roll R, Roll J-P (1999) Specific whole-body shifts induced by frequency-modulated vibrations of human plantar soles. Neurosci Lett 266(3):181–184

Kildal J (2010) 3D-press: haptic illusion of compliance when pressing on a rigid surface. In: International conference on multimodal interfaces and the workshop on machine learning for multimodal interaction, ACM, New York, pp 21

Kildal J (2012) Kooboh: variable tangible properties in a handheld haptic-illusion box. In: haptics: perception, devices, mobility, and communication, Springer, Berlin, pp 191–194

Kimura F, Yamamoto A, Higuchi T (2010) Development of a 2-Dof softness feeling display for tactile tele-presentation of deformable surfaces. In: 2010 IEEE international conference on robotics and automation, pp 1822–1827

Klatzky R, Lederman S (1999) Tactile roughness perception with a rigid link interposed between skin and surface. Percept Psychophys 61(4):591–607

Klatzky R, Lederman S, Hamilton C, Grindley M, Swendsen R (2003) Feeling textures through a probe: effects of probe and surface geometry and exploratory factors. Percept Psychophys 65(4):613

Kobayashi Y, Osaka R, Hara T, Fujimoto H (2008) How accurately people can discriminate the differences of floor materials with various elasticities. IEEE Trans Neural Syst Rehabil Eng 16(1):99–105

Kontarinis DA, Howe RD (1995) Tactile display of vibratory information in teleoperation and virtual environments. Presence Teleoper Virtual Environ 4(4):387–402

Kuchenbecker KJ, Fiene J, Niemeyer G (2006) Improving contact realism through event-based haptic feedback. IEEE Trans Visual Comput Graphics 12(2):219–230

LaMotte R (2000) Softness discrimination with a tool. J Neurophysiol 83(4):1777

Lang J, Andrews S (2011) Measurement-based modeling of contact forces and textures for haptic rendering. IEEE Trans Visual Comput Graphics 17(3):385–391

Lecuyer A, Coquillart S, Kheddar A, Richard P, Coiffet P (2000) Pseudo-haptic feedback: can isometric input devices simulate force feedback? In: Virtual reality, 2000. Proceedings. IEEE, pp 83–90

Lederman SJ, Klatzky RL (1987) Hand movements: a window into haptic object recognition. Cogn Psychol 19(3):342–368

Massimino MJ, Sheridan TB (1993) Sensory substitution for force feedback in teleoperation. Presence Teleoper Virtual Environ 2(4):344–352

Morioka M, Griffin MJ (2002) Dependence of vibrotactile thresholds on the psychophysical measurement method. Int Arch Occupational Environ Health 75(1–2):78–84

Morioka M, Whitehouse DJ, Griffin MJ (2008) Vibrotactile thresholds at the fingertip, volar forearm, large toe, and heel. Somatosens Mot Res 25(2):101–112

Okamoto S (2010) Tactile transmission system and perceptual effects of delayed tactile feedback. PhD thesis, Tohoku University

Okamura A, Dennerlein J, Howe R (1998) Vibration feedback models for virtual environments. In: 1998 IEEE international conference on robotics and automation, 1998. Proceedings, vol 1. pp 674–679

Okamura AM, Cutkosky MR, Dennerlein JT (2001) Reality-based models for vibration feedback in virtual environments. IEEE/ASME Trans Mechatron 6(3):245–252

Pense-Lheritier A-M, Guilabert C, Bueno M, Sahnoun M, Renner M (2006) Sensory evaluation of the touch of a great number of fabrics. Food Qual Prefer 17(6):482–488

Ribot-Ciscar E, Vedel J, Roll J (1989) Vibration sensitivity of slowly and rapidly adapting cutaneous mechanoreceptors in the human foot and leg. Neurosci Lett 104(1):130–135

Rust J, Keadle T, Allen D, Shalev I, Barker R (1994) Tissue softness evaluation by mechanical stylus scanning. Text Res J 64(3):163–168

Scilingo EP, Bianchi M, Grioli G, Bicchi A (2010) Rendering softness: integration of kinesthetic and cutaneous information in a haptic device. IEEE Trans Haptics 3(2):109–118

Scott-Blair G, Coppen F (1940) The subjective judgement of the elastic and plastic properties of soft bodies. Br J Psychol 31:61–79

Srinivasan M, LaMotte R (1995) Tactual discrimination of softness. J Neurophysiol 73(1):88–101

Suzuki T, Mabuchi K, Nishimura H, Saito T, Kakuta N, Kunimoto M, Shimojo M, Ishikawa M (1999) The electrical control of pressure sensations: the relationship between stimulation signals and subjective intensities and areas. In: first joint BMES/EMBS conference, pp 457–457

Takahiro Y, Okamoto S, Konyo M, Hidaka Y, Maeno T, Tadokoro S (2010) Real-time remote transmission of multiple tactile properties through master-slave robot system. In: 2010 IEEE international conference on robotics and automation, IEEE, pp 1753–1760

Tan H, Durlach N, Beauregard G, Srinivasan M (1995) Manual discrimination of compliance using active pinch grasp: The roles of force and work cues. Percept Psychophys 57(4):495–510

Tiest WMB, Kappers AM (2009) Cues for haptic perception of compliance. IEEE Trans Haptics 2(4):189–199

Vedel J, Roll J (1982) Response to pressure and vibration of slowly adapting cutaneous mechanoreceptors in the human foot. Neurosci Lett 34(3):289–294

Verrillo RT (1966) Vibrotactile thresholds for hairy skin. J Exp Psychol 72(1):47

Visell Y (2009) Tactile sensory substitution: Models for enaction in hci. Interact Comput 21 (1–2):38–53

Visell Y, Giordano BL, Millet G, Cooperstock JR (2011) Vibration influences haptic perception of surface compliance during walking. PLoS One 6(3):e17697

Chapter 4
Perception and Synthesis of Sound-Generating Materials

Bruno L. Giordano and Federico Avanzini

4.1 Introduction

The mechanical properties of sound-generating objects and events in our environment determine lawfully the acoustical structure of the signals they radiate (e.g. Fletcher and Rossing 1991). The ability of listeners to estimate the mechanical properties of everyday non-vocal, non-music sound sources based on acoustical information alone has been the object of empirical research for more than three decades (Vanderveer 1979). Given the lawful specification of the mechanics of the sound source in the acoustical structure, and the adaptive tendency to interpret sensory information in terms of the properties of objects and events in the environment, it is thus not surprising that source-perception abilities are often remarkably accurate (see Lutfi 2007, for a review).

The concept of material is central to the study of source perception from both a theoretical and empirical point of view. The theoretical relevance of this concept originates from the work of Gaver, who outlined a widely influential taxonomy of everyday sound events (Gaver 1993). Accordingly, non-vocal sound sources offer perceptual systems with information about "materials in interaction", and, at the most general level of the taxonomy, can be classified into three categories depending on the state of matter of the vibrating sound-generating substance: (i) solid sound sources (e.g. clapping); (ii) liquid sound sources (e.g. pouring coffee); (iii) aerodynamic/gaseous sound sources (e.g. wind; explosions). For a variety of reasons (e.g. easiness in manipulation of source mechanics; ecological pervasiveness), the

B.L. Giordano (✉)
Institute of Neuroscience and Psychology, University of Glasgow, 58 Hillhead Street, Glasgow
G12 8QB, UK
e-mail: brungio@gmail.com

F. Avanzini
Department of Information Engineering, University of Padova, Via Gradenigo 6/A,
35131 Padova, Italy
e-mail: avanzini@dei.unipd.it

© Springer-Verlag London 2014
M. Di Luca (ed.), *Multisensory Softness*, Springer Series on Touch and Haptic Systems,
DOI 10.1007/978-1-4471-6533-0_4

empirical study of the perception of source mechanics has favoured solid sound sources, and investigated the audition of three different attributes: (i) geometry (e.g. shape of a struck bar; Lakatos et al. 1997); (ii) material (e.g. hardness of a mallet striking a pan; Freed 1990); (iii) properties of the interaction between sound-generating objects (e.g. bouncing vs. breaking of objects; Warren and Verbrugge 1984). The empirical centrality of materials then originates quite simply from the fact that among these three perceptual abilities, material perception has been by far the most studied.

In the first part of this chapter, we will review studies on the perception of material properties from sound. We will initially describe the available empirical evidence on the perception of the state of matter, and then detail the psychophysics literature on the recognition of the material properties of stiff solid objects. Importantly, this last group of studies did not investigate highly compressible solid materials such as soft rubbers, most likely because of the often perceptually negligible acoustical energy they radiate when set into vibration. A subsequent section will detail studies of two classes of deformable materials: fabrics and liquids. Although these investigations addressed the perception of material-independent properties such as texture or liquid amounts, they are summarized here because of the potential interest to future research in the field. The last portion of this part ends with a presentation of studies on the perception and motor-behaviour effects of stiff materials in audio-haptic contexts (see Chap. 2 for visual-haptic contexts).

Studies in ecological perception are the starting point for the development of interactive sound synthesis techniques that are able to render the main perceptual correlates of material properties, based on physical models of the involved mechanical interactions. In the second part of this chapter we will then review recent literature dealing with contact sound synthesis in such fields as sonic interaction design and virtual reality. Special emphasis will be given to softness/hardness correlates in impact sounds, associated to solid object resonances excited through impulsive contact, and rendered using modal synthesis techniques. We will also summarize recent advances in terms of optimization and automation of analysis-synthesis schemes. Two final sections will address less developed literature on the sound synthesis and rendering of deformable objects (notably textiles), aggregate objects (e.g. sand, snow, gravel, and so on), and liquids.

4.2 Perception

4.2.1 State of Matter

The first study that gave some indication of the perceptual relevance of the state of matter was carried out by Ballas (1993). Participants in this study were asked to rate a set of 41 environmental sounds that included liquid, aerodynamic (gases), and solid sounds along a variety of scales meant to assess their perceptual representation (e.g. dull vs. sharp timbre) but also aspects of their cognitive processing (e.g. sound familiarity; similarity of sound to a mental stereotype). Principal component analysis of the rating data outlined three orthogonal judgment dimensions, interpreted as measuring: (i) the identifiability of the sound event (e.g. familiarity scale); (ii) the sound timbre (e.g. dull vs. sharp rating scale); (iii) attributes of the categorical representation (number of similar sounds). A cluster analysis of these principal components revealed one cluster of liquid sounds, which however also included sounds produced in a water context (e.g. boat whistle). The other three clusters included both solid and aerodynamic sources, and grouped together either signals with similar functions (e.g. cluster of signalling sounds, and cluster of door sounds), or highly transient sounds independently of whether they were generated by aerodynamic or solid events (cluster of transient sounds such as a stapler). Overall, this initial study lent some support to the hypothesis that listeners are capable of differentiating between states of matter, although the clustering structure was most likely influenced by a number of factors related to the cognitive processing and higher-order information about the sound signal such as the context where it is generated.

Gygi et al. (2007) investigated the dissimilarity ratings and free sorting of a set of 100 sounds that included living human and non-human sounds (both vocalizations and non-vocalizations), and non-living sounds generated by solid, liquid and aerodynamic sources. In the dissimilarity-ratings experiment participants were presented with all of the possible pairs of stimuli, one pair at a time, and were asked to rate how dissimilar they were. In the free-sorting experiment, participants were presented with all of the stimuli and were asked to create groups of similar sounds. Dissimilarity ratings were analysed with a multidimensional scaling (MDS) algorithm (Borg and Groenen 1997). In general, MDS models the input dissimilarities as the between-stimulus distance within an Euclidean space in which dissimilar stimuli are located further apart. The tendency of stimuli to cluster, i.e., to form tight groups within the MDS space, can thus give an indication of the ability of listeners to differentiate between different-group stimuli (see Ashby 1992, for the relationship between dissimilarity, categorization and discrimination). Overall, both the dissimilarity-ratings and free-sorting data revealed a tendency to group stimuli based on source attributes and, importantly, to differentiate between solid, liquid and aerodynamic events. The tendency to group together sounds generated with substances in the same state was more evident in the free-sorting than in the dissimilarity-ratings data. Indeed, clustering in the dissimilarity-ratings MDS space appeared to be more driven by acoustical attributes that are not always differentiated between diverse sound-generating mechanical systems (e.g. gunshots and footstep sounds were clustered together because they both comprised sequences of transient impact-like sounds). This discrepancy can be explained by the fact that dissimilarity-ratings data are more sensitive to differences in acoustical structure than free sorting data (Giordano et al. 2011).

Further support for the hypothesis of a perceptual relevance of the state of matter was obtained by Houix in an experiment on the free sorting of 60 sounds encountered in a kitchen context (Houix et al. 2012). Vocal sounds were not included in order to eliminate possible distortions in the sorting data independent of the state of matter arising from the likely strong perceptual and attentional salience of the vocal/

non-vocal distinction (Belin et al. 2000; Lewis et al. 2005; Gygi et al. 2007) and from differences between the cognitive processing of living and non-living sounds (Giordano et al. 2010). Consistently with Gygi et al. (2007), participants appeared to group together sounds based on the mechanics of the sound source even when not explicitly required to do so, and created isolated clusters of solid objects, machine and electric device sounds, liquid sounds, and aerodynamic sounds.

Overall, the studies summarized up to this point reveal that the state of matter is likely to structure the cognitive organization of everyday sound sources. These studies, however, do not give evidence concerning the actual ability to recognize the state of matter of a sound-generating substance. A number of sound-generating human-made solid objects have indeed been designed to create "state of matter" illusions (e.g. rainsticks; wind and thunder machines used for centuries in theatres). A recent study by Lemaitre and Heller (2013) addressed this point rigorously. A set of 54 sounds were generated with three different types of interaction for each state of matter (solid: friction, deformation and impact; liquid: splashing, dripping and pouring; gases: whooshes, wind and explosions). Sound duration was gated at different levels. Overall, untrained listeners were able to recognize the state of matter at above-than-chance levels across gating durations and interaction types (75 % correct). Figure 4.1 shows the spectrogram of a set of sound stimuli used in their study.

4.2.2 Perception of Stiff Solid Materials

Except for the recent study by Lemaitre and Heller (2012), investigations into the auditory perception of stiff solid materials were all carried out with isolated impact sounds. In particular, real or simulated impact sounds were generated by the inter-action between two objects: the hammer and the sounding object, the former being much more damped than the latter (e.g. when a drum stick strikes a cymbal, the impact sets into vibration the drum stick for a much shorter time than the cymbal). Given the high damping of the former, the sound signal presented to the listeners in these studies contains little or no acoustical energy radiating directly from the ham-mer, and can be assumed to be the product of acoustical radiation from the sounding object alone. Notably, however, the material properties of the hammer still influence the acoustical structure of the radiated sound signal (Fletcher and Rossing 1991). For example, stiffer hammer materials produce a decrease in the duration of the contact between the hammer and sounding object during the impact, resulting in a more efficient excitation of the high-frequency vibrational modes of the sounding object and, consequently, in an increase in the high-frequency energy of the radiated sound.

Impact sounds can be modeled as the sum of sinusoids whose amplitude decays exponentially starting from the onset of the sound signal. Ignoring perceptually neg-ligible delays, the temporal location of the sound onset essentially corresponds to the time of contact between the hammer and the sounding object. The material proper-ties of the sounding object (elastic coefficients; density), together with its geometry,

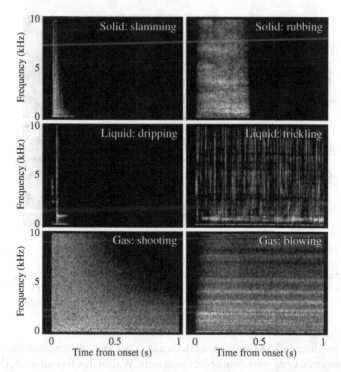

Fig. 4.1 Spectrum of the sound stimuli generated with solid, liquid, and gaseous substances. The *left panels* group impulse-like sounds, generated as a consequence of a temporally localized input of energy in the sound-generating system. The *right panels* show continuous sounds, generated through a temporally distributed input of energy into the sound-generating system. Level from *black* (low) to *white* (high). Data courtesy of Lemaitre and Heller (2013)

determine the frequency of the exponentially decaying spectral components: stiffer and denser materials (and smaller object sizes) produce higher frequency spectral components. The material properties of the hammer, together with the geometry of the hammer and the properties of the hammer-sounding object interaction (e.g. striking force; duration of the hammer/sounding object contact) determine the initial amplitude of the spectral components, and the overall spectral distribution of energy. Stiffer hammers determine higher energy levels in the high-frequency regions. Importantly, the material properties of the sounding object also determine the decay times of the spectral components. Overall, stiffer materials produce spectral components characterised by a slower decay. The velocity of the decay of the spectral components is, however, not constant across all of the spectral components: to a rough approximation, higher-frequency components decay faster than low-frequency components. Wildes and Richards (1988) outlined a simplified yet widely influential model of the relationship between spectral frequency and energy decay. In their model, an increase in spectral frequency produces a linear increase in the decay time of the spectral components. In particular, for stiffer materials the increase in spectral

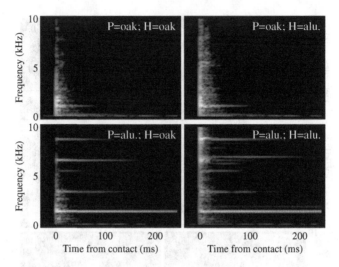

Fig. 4.2 Detail of the spectrotemporal structure of impact sounds generated by striking a 450 cm^2 square plates (P) made of oak or aluminium with an aluminium or oak hammer (H). Level from low (*black*) to high (*white*). Data from Giordano et al. (2010)

frequency produces a less pronounced decrease in decay time, whereas the contrary is true for more flexible sounding object materials. Within this formulation, the slope of the linear relation between spectral frequencies and decay time is assumed to be a reliable acoustical measure of the stiffness of the sounding object. This variable takes the name of tan φ, and has been assumed to measure a geometry-invariant acoustical correlate of the sounding-object material. Figure 4.2 displays the time-varying spectrum of the sounds generated by striking an aluminium and oak square plate (area = 450 cm^2; thickness = 1 cm) with a semi-spherical aluminium or oak hammer. Note: (i) the longer duration, i.e., lower decay time of the spectral components in the aluminium plate; (ii) the steeper decrease of decay times with frequency for the oak plate; (iii) the higher energy of high-frequency components for the sounds generated with the aluminium hammer.

Two of the earliest studies on the identification of the material of a sounding object were carried out with real sounds, and were primarily concerned with ascertaining identification performance rather than the acoustical factors involved in the identification process. Gaver (1988) struck wood and steel bars of different lengths with a rubber hammer. The vibration of the bars was externally damped with a carpet, on top of which they rested while being struck. Participants were presented with recorded impact sounds, and reached nearly perfect material-identification performance. Kunkler-Peck and Turvey (2000) investigated the ability to identify simultaneously the material and shape of a freely-vibrating plate (sounding object) struck with a steel pendulum (hammer). All of the plates had the same area, but differed in both shape (circle, triangle, rectangle) and material (steel, wood, Plexiglas). Sounds were generated live by the experimenter, while being occluded from the view of the experiment participant. The identification of material was virtually perfect (only one misidentification was recorded) and was more accurate than the identification of shape which, however, was significantly better than chance.

Another set of early studies on the perception of the material properties of sounding object focused on synthetic sounds (Avanzini and Rocchesso 2001; Lutfi and Oh 1997; Klatzky et al. 2000). Stimuli in these studies were synthesized as the sum of exponentially decaying sinusoids. In the studies by Avanzini and Rocchesso (2001) and by Klatzky et al. (2000), material properties were controlled by manipulating tan ϕ. Participants in these studies were either asked to identify sounding-object materials (Avanzini and Rocchesso 2001; Klatzky et al. 2000), or to discriminate the change in material between two subsequently presented sounds (Lutfi and Oh 1997). Based on the work by Wildes and Richards (1988), participants' responses were hypothesized to focus on the acoustical information contained in the decay of the spectral components, because it would specify materials more reliably than their frequency. In practice, however, participants' responses were influenced by both decay and frequency information. In the study by Lutfi and Oh (1997) the reliance of the responses of some participants on frequency information was so strong that it effectively overshadowed the perceptual effect of decay.

The empirical observation that the auditory perception of materials can be strongly influenced by sound frequency was problematic because it contradicted the theoretical framework that dominated the field, the ecological approach to perception (Gibson 1966, 1979; Michaels and Carello 1981). Accordingly, it was thought that source perception would rely on the detection of invariants, i.e., parameters of the acoustical structure that specify reliably the source property under judgment independently of variations in non-target source properties (e.g. an invariant for sounding object material specifies this property reliably and independently of changes in nonmaterial properties such as size). Since acoustical parameters such as tan ϕ were thought to constitute invariant information about materials, it was surprising that the identification of material was also influenced by sound frequency, a variable that is influenced by both the material and geometry of the sounding object (Fletcher and Rossing 1991). Carello et al. (2003) argued that the focus of listeners on frequency information was an artefactual product of the synthetic nature of the sound signals which provided participants with impoverished material-related information. Giordano and McAdams (2006; Giordano 2003) addressed this issue within a materialidentification study conducted with real sounds recorded by striking wood, plastic, metal and glass plates of different sizes. Importantly, an analysis of the acoustical structure revealed that sounds were rich in information that differentiated between materials independently of variations in size. Among them was a psychoacoustically plausible derivation of the tan ϕ coefficient. When asked to identify materials, however, participants did not appear to fully exploit this invariant acoustical information. Indeed, when it came to differentiating between gross material categories, i.e., wood and plastic on the one hand, and metal and glass on the other, identification performance was perfect and could be accounted for by a focus on various acoustical features among which tan ϕ. However, identification performance within each of these two categories was virtually at chance level because participants differentiated between metal and glass or between wood and plastic by relying exclusively on the size of the plate or, from the acoustical point of view, on the lowest spectral component. Identification confusions between wood and plastic were also observed by Tucker and Brown (2003) in a study on real sounds, and were instead inconsistent with the results by Kunkler-Peck and Turvey (2000) obtained with sounds generated live. It is unclear whether participants in the latter study were exposed to additional material-related information through eventual uncontrolled sounds generated while the experimenter hung the plates on the apparatus used to strike them.

The effect of frequency information on identification responses in the study by Giordano and McAdams (2006) was consistent with previous studies of synthetic sounds, and, together with the observation of the availability of acoustical information for the size-independent identification of material, disconfirmed the impoverished information hypothesis for the perceptual focus on sound frequency. Various hypotheses can be advanced to explain the influence of object size and sound frequency on the identification of the material of an object. It might be for example argued that participants in the study by Giordano and McAdams erroneously interpreted all of the available acoustical information in terms of object material because they were not informed of the variation in the size of the plates. This hypothesis would, however, not explain why participants in this study identified small glass and metal plates, which generated higher-frequency sounds, as being made of glass because metal is denser and stiffer than glass (Waterman and Ashby 1997) and should thus generate higher-frequency sounds. Another possible interpretation for the reliance of material identification on sound frequency is more subtle. Studies on the perception of musical timbre show that the influence of pitch on judgments of the dissimilarity of musical timbres grows with the range of variation of pitch within the experimental set and, in particular, becomes relevant at the expense of a focus on non-pitch acoustical information when participants are exposed to pitch variations larger than one octave (Handel and Erickson 2001; Marozeau et al. 2003; Steele and Williams 2006). Accordingly, the reliance of material identification on frequency observed in previous studies was determined by the comparatively large variation of this pitch-related acoustical variable (e.g. six octaves in Giordano and McAdams 2006). Consistently with this interpretation, a subsequent material-identification experiment by McAdams et al. (2010), carried out with a sound set that included a much smaller frequency variation (less than half an octave) revealed no effect of this variable on identification responses (see McAdams et al. 2004; Giordano 2005; McAdams et al. 2010 for dissimilarity-ratings studies of impacted sound sources). A final plausible explanation of the effect of frequency on identification responses is that it is the product of the internalization of a statistical regularity in the acoustical environment. Accordingly, listeners identified the high-pitched sound of a small metal plate as being made of glass because of the small size of everyday freely vibrating glass objects (e.g. clinking glasses).

Overall, studies on the identification of the material of impacted sounding objects reveal a nearly perfect ability to differentiate between gross categories of materials (wood or plastic vs. metals or glass), and a number of frequency-dependent biases in the identification of materials within these categories. From the acoustical point of view, the ability to differentiate between these gross categories appears to be well explained by the perceptual processing of an acoustical measure of the damping of vibrations, tan φ (Giordano and McAdams 2006; McAdams et al. 2010; Avanzini and Rocchesso 2001; Klatzky et al. 2000). Notably, however, this acoustical variable might not account well for the differentiation of gross material categories in non-impact interaction types (Lemaitre and Heller 2012). From the mechanical point of view, the ability to differentiate between gross material categories appears to be robust to variations in the geometry of the struck object (see also Ren et al. 2013), and vulnerable only to the external damping of the vibration of the sounding object, either when it is submerged underwater (cf., identification of material in sounds recorded under water; Tucker and Brown 2003), or when it is attached to a dampening soft-plastic surface (Giordano 2003).

The studies reviewed up to this point investigated the identification or discrimination of material categories, but did not assess the ability to estimate quantitative material attributes such as their hardness/softness. Giordano et al. (2010) assessed the ability of participants to discriminate and rate the hardness of sounding objects in the presence or absence of training. Sounds were generated by striking variable size plates of different materials with hammers of different materials. In an initial experiment, listeners received correctness feedback when asked to discriminate the hardness of sounding objects. Within a limited number of blocks trials they were able to quickly learn to discriminate the hardness of sounding objects independently of variations in its size, and independently of variations in the hardness of the hammer. The training received in the discrimination experiment generalized to a second hardness-rating experiment where they did not receive correctness feedback. Also in this experiment they were able to accurately rate the hardness of sounding objects independently of variations in their size and of variations in the material of the hammer. Importantly, another group of participants who did not receive prior discrimination training appeared to estimate the hardness of sounding objects by focusing on the target mechanical properties, but did not ignore the hardness of the hammer and the size of the sounding object which still influenced their rating responses although to a lesser extent. Similar effects were obtained with another group of untrained listeners who rated the hardness of sounding objects when presented with synthetic impact sounds. Again, their estimates were most strongly influenced by the synthetic parameter modeling the hardness of the sounding object, but they were influenced, to a lesser extent, by the frequency of the sounds, and by the acoustical parameters that in real sounds were most strongly influenced by the impact properties. Overall, the study by Giordano et al. (2010) confirms the tendency of untrained listeners to estimate the material properties of the sounding object by considering also non-material parameters such as the size of the sounding objects. The fact that trained listeners are able to estimate the hardness of sounding objects independently of their size confirms the presence of perceptually available yet not completely exploited information for accurate material perception.

Only three studies investigated the ability to perceive the material properties of the hammer (Freed 1990; Lutfi and Liu 2007; Giordano et al. 2010). In the study by Freed (1990), participants were presented with the sound of variable-sized metallic pans struck with mallets of different hardness and were asked to estimate the hardness of the hammer. Hardness estimates appeared to be accurate, and independent of the size of the sounding object. Lutfi and Liu (2007) investigated the weighting of the amplitude information of the spectral components of synthetic impact sounds within a hammer-hardness discrimination task, and assessed the extent to which the weighting strategies are reliable across different days. Although different individuals were characterized by largely diverse patterns of information weighting, the weighting strategies of each single individual were highly replicable.

The study by Freed lent support to the hypothesis that participants are extremely accurate at perceiving the material properties of hammers. Combined with the results concerning the often imperfect identification of the material of sounding objects, it would thus appear, paradoxically, that the auditory system is better equipped at detecting the properties of an object that only indirectly structures the acoustical signal through its effects on the sounding object vibration—the hammer—rather than at detecting the material properties of the sound-radiating sounding object. However, it should be noted that in Freed (1990) the material of the hammer varied while that of the sounding object was kept constant. As such, this study offered only a rather limited test of the extent to which the auditory perception of hammer materials is truly invariant. In a follow-up study carried out with both real and synthetic sounds, Giordano et al. (2010) asked participants to estimate the hardness of hammers that struck plates of different size and material. Across multiple conditions, participants were able to estimate accurately the hardness of the hammer only when receiving trial-by-trial feedback on discrimination performance. In the absence of such a feedback, they appeared instead to estimate hammer hardness based on properties of the hammer-sounding object impact, such as the duration of their contact. Importantly, impact properties are influenced by both the material of the hammer and of the sounding object, i.e., perception of the hammer material relied on a less-than optimal mechanical variable.

Overall, studies on the perception of solid materials have revealed that in the absence of explicit training or feedback on the correctness of their responses, the perceptual abilities that listeners bring into the experimental context are often less than perfect, and are influenced by non-target mechanical properties of the sound source. These results make it rather unlikely that naive listeners rely on acoustical invariants that specify accurately a target source property. In line with this interpretation, it has been frequently observed that listeners perceive sound source properties by relying on multiple attributes of the acoustical signal:

- the perception of the geometrical properties of the sounding object is influenced by the frequency of the spectral components (Lakatos et al. 1997; Lutfi 2001; Houix 2003), by the properties of the sound decay (Lutfi 2001; Houix 2003), and by the distribution of energy across the spectrum, as measured by the spectral centroid, the amplitude-weighted average of the spectral frequencies (Lakatos et al. 1997);
- the perception of the material properties of the sounding object is influenced by the properties of the sound decay (Lutfi and Oh 1997; Klatzky et al. 2000; Avanzini and Rocchesso 2001; Giordano and McAdams 2006; McAdams et al. 2010), but also

by sound frequency (Klatzky et al. 2000; Avanzini and Rocchesso 2001; Giordano and McAdams 2006);

• the perception of the material and mass of the hammer is influenced by both loudness and spectral centroid (Freed 1990; Grassi 2005).

The empirical observation of the reliance of perceptual judgment on multiple features of the sound signal opens up the question of how listeners establish their perceptual weight, i.e., the strength of their influence on the perceptual estimation of the source property. Giordano et al. (2010) investigated the extent to which the perceptual weight of an acoustical feature could be accounted for by two different principles. Firstly, the accuracy of the acoustical information, i.e., the extent to which an acoustical feature specifies accurately the mechanical property. Secondly, the "exploitability" of the acoustical information, i.e., the extent to which the perceptual system can use the available acoustical information given limitations in discrimination, learning and memory.

The information-accuracy principle is a quantitative extension of the invariant-information hypothesis originating within the ecological approach to perception. Accordingly, an ideal observer that carries out a source-perception task (e.g. ratings of the hardness of sounding objects) can achieve a different performance level when focusing on different acoustical features. A given acoustical feature is thus characterized by a specific task-dependent accuracy score, i.e., it affords a given performance level that ranges from chance level to perfect (e.g. zero to perfect correlation between hardness ratings and actual hardness levels). Giordano et al. (2010) measured such task-dependent information-accuracy scores by analysing a large database of impacted sounds, and hypothesized an increase of perceptual weight with an increase in their value. The information-exploitability principle states instead that, independently of the task at hand, the perceptual response will be more strongly influenced by acoustical features that, in general terms, are processed more efficiently by the observer. For example, acoustical features that are better discriminated will have a stronger influence on the estimation of a given source property (e.g. Ernst and Banks 2002). Similarly, perceptual weights will be higher for acoustical features that observers learn more quickly to associate with a given source property, or whose association with a source property is stored in memory more stably. In Giordano et al. (2010), information exploitability was measured by the ability of listeners to retain and generalize the perceptual focus on a given acoustical features from a condition where they received trial-by-trial feedback on response correctness to a subsequent condition where such feedback was not available.

In an initial discrimination experiment, Giordano et al. (2010) observed that listeners learn quickly to discriminate the hardness of sounding objects, whereas they require a longer training to reach the same target performance level when discriminating the hardness of hammers. When they received trial-by-trial performance feedback participants focused on the most accurate acoustical features for the discrimination of the hardness of both objects. However, in the absence of such a feedback the same participants in a second hardness-rating experiment were able to retain the focus on the most accurate information only when it came to estimating the hardness of sounding objects, but not of hammers. As such, the perceptual focus on accurate information appeared to be strongly limited by the ability of participants to learn and retain in memory perceptual criteria acquired while carrying out the initial discrimination task. Overall, this study thus shows that the weighting of acoustical information for the perception of sound sources is not always dominated by the accuracy of acoustical information, but is also determined by task-independent limitations in the processing abilities of the perceptual system. This view is consistent with the observation by Lutfi and Stoelinga (2010) that performance in the perception of the properties of a struck bar can be accounted for by the ability of listeners to discriminate the features of the sound signal.

4.2.3 Comparison of Material and Interaction Perception

Everyday non-vocal sound sources can be differentiated based on both the sound-generating materials and the type of interaction that sets them into vibration: solid materials can be set into vibration by plastic deformations (e.g. crumpling paper), impacts, scraping and rolling; liquid interactions include dripping, pouring, and splashing; interactions for aerodynamic/gaseous sound sources include explosions, gusts, and wind-like turbulence (Gaver 1993). Given the centrality of the construct of interaction to the organization of everyday sound sources, and the strong effects they have on the structure of the acoustical signal (see Figs. 4.1 and 4.3), it is thus natural to ask which, among materials and interactions, are central to the cognitive

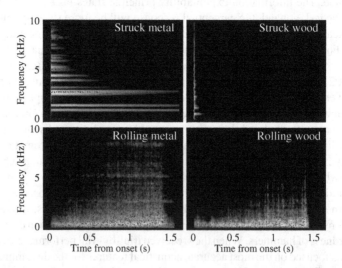

Fig. 4.3 Spectrum of the sound stimuli generated by hitting or rolling the same metal (*left panels*) and wood (*right panels*) cylinders. Level from low (*black*) to high (*white*). Data courtesy of Lemaitre and Heller (2012)

organization of everyday non-vocal sounds, and which our auditory system can process better. Lemaitre and Heller addressed these questions in two recent studies (Lemaitre and Heller 2012, 2013).

In Lemaitre and Heller (2013), participants were presented with sounds generated by setting into vibration various substances (solids, gases and liquids) with different types of actions. They carried out a label-verification experiment, i.e., they had to decide whether a given label was an appropriate description of the sound-generating event. Labels could belong to one of five categories: (i) state of matter; (ii) interaction type (e.g. friction for scraping and squeaking or deformation for tearing and crumpling); (iii) specific interaction (e.g. scraping); (iv) manner of action (e.g. scraping rapidly); (v) object of the action (e.g. scraping a board). Labels belonging to the specific-interaction category were verified more quickly and accurately than labels from any other category, suggesting a central role of this description level in the cognitive organization of the sound stimuli. The same conclusion was supported by the results of a second priming experiment where sounds primed a lexical decision task carried out on subsequently presented labels belonging to either the interaction type or specific interaction category. Responses were faster and more accurate for specific-interaction labels.

In Lemaitre and Heller (2012), sounds were generated by applying different interaction types (rolling, scraping, hitting and bouncing) to hollow cylinders made of four different materials (plastic, wood, metal and glass). In an initial experiment, participants rated how well the sound conveyed either a specific interaction type or a specific material. Performance measures derived from the ratings data were significantly better when participants judged the interaction type. In a second experiment, participants were asked whether a target sound had been generated with a given interaction type or material described by a label presented before the sound onset. Reaction times were faster for the identification of the interaction type.

Overall, the studies by Lemaitre and Heller support the hypothesis that the manner in which substances and objects are set into vibration play a more central role in the cognitive organization of non-vocal everyday sounds. Further, interaction types appear to be associated with acoustical fingerprints that are processed more quickly by the auditory system than those that characterize materials and states of matter.

4.2.4 Perception of Deformable Materials

Sounds produced by fabrics are the result of complex interactions that include sliding friction but also crumpling sounds due to buckling of the fabric on itself. The frictional component of these sounds is largely influenced by the texture of the fabric, a parameter mostly related to the fine-grained geometry of the object rather than to geometry-independent material properties. The reader is referred to the work by Lederman and co-workers on the audio-haptic perception of texture (Lederman 1979; Lederman and Klatzky 2004).

In the scientific literature there appears to be a lack of studies on the auditory perception of material-related properties of fabrics and textiles. On the other hand, the definition of objectively measurable properties of fabrics (the term Fabric Objective Measurement—FOM—is often used) is a central issue in the field of textile and apparel industry. For this reason, the perception of fabrics and textiles has been studied in the context of the evaluation of fabric "hand", quality, and related performance attributes. Judgments of hand feel properties of fabric are traditionally given by panels of experts, and sets of hand feel properties and corresponding scales have been defined (Civille and Dus 1990), most of which are related to the tactile sense. Among them, surface softness (on a subjective scale *soft* → *hard*) and, interestingly, two auditory properties: "noise intensity" (*soft* → *loud*), and "noise pitch" (*low/bass* → *high/sharp*). The Kawabata evaluation system (KES) (Kawabata 1980) has been developed under the assumption that the main characteristics of fabric responsible for hand feel depend on many physical properties, including dimensional changes at small forces (tensile, shear, compression, bending), surface properties (roughness and friction), and surface thermal insulation. KES testing instruments estimate various mechanical properties of fabric, which are then related to sensory signals acquired through hand-contact.

Within this research area, the role of auditory information in the subjective evaluation of fabrics has been assessed in a series of studies on the effects of sound on hand-feel properties (Cho et al. 2001, 2005). In Cho et al. (2001), a set of psychophysical experiments was presented concerning the characteristics of frictional sounds produced by interaction with fabrics, and their effect on the perceived quality of fabrics. Among various fabric-sound features, the sound "level range" ΔL was found to exhibit a positive correlation with perceived softness and pleasantness. In Cho et al. (2005), the reaction of observers to frictional sounds of warp-knitted fabrics was measured in terms of physiological responses (EEG, respiration rate, skin conductance level, etc.). In this study, the psychoacoustical measures of roughness and fluctuation strength were strongly correlated with the perceived pleasantness of fabric sounds.

Although liquid sound sources have been investigated in a number of studies, few of them assessed the ability of listeners to perceive a mechanical property of the sound source. The most active area of research on liquid sounds is indeed of an applied nature, and aims for example to assess how they can be used to mask road-traffic noise (e.g. De Coensel et al. 2011; Jeon et al. 2012). Overall, this research field shows promising potentials of water sounds for the improvement of the perceptual quality of urban soundscapes. Consistently, a semantic-differential study of environmental sounds carried out by Kidd and Watson (2003), revealed that across a large set of environmental sounds, liquid sounds are among the least harsh (e.g. splashing vs. breaking light bulb sound) and the most appealing (e.g. waterfall vs. scraping wood sounds).

Although not directly related to the auditory perception of the mechanics of the sound source, the study carried out by Geffen et al. (2011) gives interesting indications on how liquid sounds might be encoded in the auditory system. It is known that the acoustical structure of natural sounds exhibit scale-invariant or self-similarity traits, as measured by the $1/f$ distribution of the power spectrum of amplitude fluctuations across different frequencies f (Voss and Clarke 1975; Attias and Schreiner 1997). Confirming these earlier studies, Geffen et al. (2011) observed that natural water sounds exhibit the same scale-invariant $1/f$ spectrum. More importantly, they observed that natural water sounds whose scale was modified simply by altering their playback speed were still perceived as natural and water like and that synthetic stimuli given by the overlap of temporally and spectrally distributed chirps were still perceived as natural and water like only when they exhibited the same scale-invariant structure and a $1/f$ spectrum.

Four studies examined the perceptual estimation of the mechanical properties of liquid sound sources. Jansson (Jansson 1993; Jansson et al. 2006), carried out a series of experiments to assess the estimation of the amount of liquids inside a shaken vessel. In Jansson (1993), participants estimated the amount of liquid in a shaken opaque container in haptic, auditory or visual conditions. Although auditory and visual estimates of the amount of liquid scaled to the actual amount in all conditions, the most accurate estimates were given when haptic information was available. Interestingly, accuracy in all of these conditions improved dramatically after participants were exposed to a prior multisensory condition where information from all of the three modalities was available, revealing a rapid calibration of the processing of information within each of the modalities. In Jansson et al. (2006), participants estimated the amount of liquid or of a solid substance held within a shaken vessel in various haptic conditions, each characterized by a different constraint on the exploratory movement that participants could execute, and in a trimodal visual-auditory-haptic condition. Consistent with the previous study, accuracy was higher in the trimodal condition and in the haptic condition where participants were allowed to shake the vessel, as opposed to only lift it. Cabe and Pittenger (2000), investigated the ability to perceive the filling level in water container. An initial auditory-only experiment revealed that listeners can accurately differentiate between liquid-pouring events where the overall level of water decreases, increases or remains constant. In a second experiment, participants were asked to fill a container up to a specified level. Accuracy was higher when participants had access to haptic, visual and auditory information as compared to when they had access to auditory information only. In a final experiment, inspired by studies on the visual estimation of time to contact, listeners were found able to predict accurately the time it would have taken for a vessel to fill completely after having heard the sound generated by filling it at various below-brim levels. Finally, Velasco et al. (2014) investigated the perception of the temperature of poured liquids. In an initial experiment, sounds were generated by pouring cold or hot water in one of four different containers (glass, plastic, ceramic and paper). Participants were very accurate at identifying whether the poured water was cold or hot for all containers. Good recognition abilities were confirmed in a second experiment that measured the implicit association between cold and hot sounds, and the "cold drink" and "hot drink" verbal labels. In a final experiment, participants rated the temperature of one cold and one hot liquid pouring sound, and of a manipulated version of the cold sound that increased high-frequency energy, and of the hot sound that increased low-frequency energy. Temperature ratings increased from the manipulated cold to the original cold to the original hot to the manipulated hot sound, suggesting an increase in perceived liquid temperature with a decrease in the low-frequency energy of the sound signal.

4.2.5 Audio-haptic Perception of Materials

The studies reviewed up to this point on the perception of stiff impacted materials reveal that geometry-independent material information is highly relevant to the auditory perception of objects and events. In contrast, it is a largely open question whether auditory material-related information is perceptually relevant when presented in a multisensory context because relatively few studies addressed this question. Chapter 2 by Klatzky and Wu in this volume reviews studies on the integration of audio-visual information about materials. Here, we review studies on the audio-haptic processing of material information. They can be divided in two groups according to the exploratory gesture used by observers to generate the sound signals: hitting and walking.

The audio-haptic perception of struck materials was investigated in three studies (DiFranco et al. 1997; Avanzini and Crosato 2006b; Giordano et al. 2010). DiFranco et al. (1997) investigated the ranking of the stiffness of simulated haptic surface when presented along with recorded sounds generated by striking materials of different stiffness. Stiffness rankings increased with the actual stiffness of the auditory or haptic objects when they were presented along with a constant haptic stiffness or when presented alone, respectively. Two groups of observers participated in the main experiment. They had either taken part in a previous experiment on the ranking of haptic stiffness (expert observers) or not (naive observers). Naive observers appeared to be more strongly influenced by auditory stiffness than expert observers (accuracy in ranking of haptic stiffness across sound-stiffness levels = 44 and 73 % correct, respectively). Notably, however, also the expert observers appeared to take into account auditory stiffness to some extent because their performance in the ranking of haptic stiffness appeared to decrease relative to what was observed when they were not exposed to simultaneous sound stimuli (83 % correct in the haptic-only condition). Avanzini and Crosato (2006b) investigated the perceptual effectiveness of a haptic-synthesis engine coupled with a real-time engine for the synthesis of impact sounds. The sound-synthesis engine in this study allowed the manipulation of the force-stiffness coefficient, a mechanical parameter influenced by the stiffness of both the hammer and sounding object that influences primarily the perceived hardness of the hammer but also of the sounding object (Giordano et al. 2010). Variable-stiffness sounds were presented along with a simulated haptic surface of constant stiffness. Consistently with what observed by DiFranco et al. (1997) with real sounds, the ratings for the stiffness of these audio-haptic events increased with the auditory stiffness.

The same model investigated in Avanzini and Crosato (2006b) was adopted by Giordano et al. (2010) in a study on the effects of audio-haptic stiffness on the motor control of striking velocity. During the initial phase of each trial, participants received continuous feedback on whether their striking velocity was within a target range. After they reached a given performance criterion, feedback was removed during a subsequent adaptation phase. During a final change phase, the audio or haptic stiffness was modified from the baseline value. Three groups of individuals (non-musicians, non-percussionist musicians and percussionists) participated in four experimental conditions: (i) auditory only; (ii) haptic only; (iii) audio-haptic congruent, where the audio and haptic stiffness was changed in the same direction (e.g. increase in both); (iv) audio-haptic incongruent. Overall, an increase in audio-haptic stiffness led to a decrease in the striking velocity during the change phase of each trial. Notably, however, whereas both non-musicians and musicians decreased striking velocity for an increase in haptic stiffness, percussionists had the exact opposite weighting of this variable, and struck stiffer haptic objects faster. In the audio-haptic condition, the control of striking velocity appeared to be dominated by changes in haptic stiffness. Also, congruency modulated the motor effect of audio stiffness, which had a significant effect only during the audio-haptic congruent condition, whereas it did not modulate the motor effects of haptic stiffness.

Three recent studies investigated the perceptual and motor effects of the properties of walked-upon materials (Giordano et al. 2012; Turchet et al. 2014; Turchet and Serafin 2014). Giordano et al. (2012) carried out an experiment on the non-visual identification of real walking grounds in audio-haptic, haptic, kinaesthetic and auditory conditions. Eight ground materials were investigated: four solid materials (vinyl, wood, ceramic and marble) and four aggregate materials (gravels of four different sizes; see Fig. 4.4 for example waveforms and spectrograms of the sound stimuli in these studies). Three of the experimental conditions were interactive, i.e., participants carried out the identification task after walking blindfolded on the ground material. In the audio-haptic condition, they had access to all of the available non-visual information. In the haptic and kinaesthetic conditions, auditory information was suppressed by means of a masking noise reproduced over wireless headphones. In the kinaesthetic condition, tactile information about ground materials was suppressed by reproducing a tactile masker through a recoil-type actuator installed in an outer sole strapped under the shoe. In the auditory condition participants did not walk on the ground materials, and heard the walking sounds they had generated during the audio-haptic condition. Given the large differences between the vibratory signals generated while walking on solid vs. aggregate materials (see Fig. 4.4), it is not surprising that in all sensory conditions participants almost perfectly discriminated between these two classes of walking grounds. Within each of these categories, identification performance varied across experimental conditions, and was maximized when participants had access to tactile information in the haptic and audio-haptic conditions. In particular, tactile information appeared to be critical for the identification of solid materials because when it was suppressed during the kinaesthetic condition identification was at chance level. More interesting were the results of the analysis of the dominance of the different sensory modalities during the audio-haptic condition. For solid materials, some evidence emerged concerning the dominance of haptic information, i.e., the sensory modality that allowed the

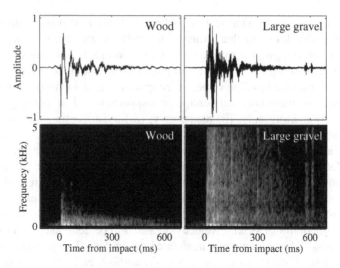

Fig. 4.4 Waveform (*top panels*) and spectrum (*bottom panels*) of a footstep sound generated while walking on wood or on large gravel, an aggregate material. Level from low (*black*) to high (*white*). Time from impact estimated approximately based on waveform. Amplitude scaled to maximum for display purposes. Data from Giordano et al. (2012)

best identification performance appeared to dominate the audio-haptic identification of solid grounds. Surprisingly, the identification of aggregate materials during the audio-haptic conditions appeared instead to be dominated by kinaesthetic information, i.e., participants focused on the worst performing sensory modality. This result was interpreted as revealing a bias in the weighting of modality-specific information: when walking on unstable grounds, such as a gravel, participants were likely not concerned with discriminating accurately the size of the gravel, but were instead concerned with keeping a stable posture by focusing on the sensory modality that would have most promptly signalled a potentially unstable posture, kinaesthesia.

Turchet and Serafin (2014) carried out a study on the congruence of simulated auditory and haptic walking ground. Participants rated the congruence of auditory and haptic materials presented simultaneously either in an active-walking condition or in a passive conditions during which they experienced the audio-haptic display while sitting on a chair. Audio-haptic congruence appeared to be maximized when both materials belonged to either the solids or aggregate category, thus confirming the perceptual relevance of the same distinction observed by Giordano et al. with real grounds. Turchet et al. (2014) finally assessed the extent to which auditory information about ground materials influences the kinematics of locomotion. Participants were instructed to walk using their normal pace on an asphalt ground. Importantly, participants could hear either the sounds they generated while walking on asphalt, or synthetic walking sounds generated in real time with a model meant to simulate wood, gravel, and snow-covered grounds. Auditory information about materials appeared to affect a number of variables related to the pace. For example, pace appeared to be slower on aggregate than on solid materials (synthetic snow and gravel vs. synthetic wood and real asphalt ground).

Overall, auditory material-related information influences both the perception of materials in audio-haptic contexts and the kinematics of sound-generating movements (see also Castiello et al. (2010), for effects of sound information on the kinematics of grasping). The available experimental evidence appears however to show that within an audio-haptic context auditory materials have weaker effects than haptic materials on both perceptual judgment and motor behaviour.

4.3 Synthesis

In light of the discussion developed in the first part of this chapter, it can be stated that appropriate sound synthesis techniques for the rendering of auditory correlates of material-related sound source properties must possess two main qualities: (i) they have to provide access to sound control parameters that can be related to ecological properties of the simulated sound-generating phenomena, and (ii) they have to be usable in real-time interactive settings, responding naturally to user actions.

Recent literature in virtual reality and sonic interaction design (Rocchesso 2011; Franinović and Serafin 2013) has explored the use of "physically based sound modeling" techniques to develop interactive sound synthesis schemes. This term refers to a set of synthesis algorithms that are based on a description of the physical phenomena involved in sound generation, whereas earlier techniques are based on a description of the sound signal (e.g. in terms of its waveform or its spectrum) and make no assumptions on the sound generation mechanisms.

Since physically based models generate sound from computational structures that respond to physical input parameters, they automatically incorporate complex responsive acoustic behaviours. A second advantage is interactivity and ease in associating motion to sound control. As an example, the parameters needed to characterize impact sounds (e.g. relative normal velocity), are computed in a VR physical simulation engine and can be directly mapped into control parameters, producing a natural response of the auditory feedback to user gestures and actions. Finally, physically based sound models can in principle allow the creation of dynamic virtual environments in which sound-rendering attributes are incorporated into data structures that provide multimodal encoding of object properties (shape, material, elasticity, texture, mass, etc.). In this way, a unified description of the physical properties of an object can be used to control the visual, haptic, and audio rendering (Avanzini and Crosato 2006b; Sreng et al. 2007).

4.3.1 Modal Sound Synthesis

Various physically based modeling techniques exist in the literature, particularly for musical instruments (see e.g. Smith 2004; Välimäki et al. 2006, for extensive reviews). A full physical simulation entails the numerical resolution of a set of partial differential equations describing mechanical and/or fluid-dynamic oscillations, as well as sound radiation in space. Finite-difference and finite-element models have been used to simulate musical instruments using this approach (Bilbao 2009). In particular, Chaigne and co-workers have developed accurate models for sound-producing mechanisms involving impacted resonators (strings, bars, plates) (Doutaut et al. 1998; Lambourg et al. 2001). A similar approach has been proposed also in the context of non-musical sound synthesis (O'Brien et al. 2002): finite-element simulations are employed for the generation of both animated video and audio. Complex audio-visual scenes can be simulated, but heavy computational loads still prevent real-time rendering and the use of these methods in interactive applications.

A more efficient technique is modal sound synthesis (Adrien 1991). Starting with the studies by van den Doel and co-workers (van den Doel and Pai 1998; van den Doel et al. 2001), this has become the most used approach for the simulation of non-musical sounds produced by mechanical contact of solid objects. Consider a resonating object described as a network of N masses connected with linear springs:

$$M\ddot{y}(t) + Ky(t) = f_{\text{ext}}(t), \qquad (4.1)$$

where y is a vector containing the displacements of the N points of the network, the mass matrix M is typically diagonal, while the stiffness matrix K is in general not diagonal because the points are coupled through springs. The homogeneous equation ($f_{\text{ext}} \equiv 0$) has in general N modal solutions of the form $y(t) = s \cdot \sin(\omega t + \phi)$, where the vector s of the *modal shapes* is an eigenvector of the matrix $M^{-1}K$ with associated eigenvalue ω^2. The eigenvectors are orthogonal with respect to the mass and the stiffness matrix, and their associated matrix $S = [s_1|s_2|\ldots|s_N]$ defines a change of spatial coordinates that transforms system (4.1) into a set of N uncoupled oscillators:

$$M_q\ddot{q} + K_q q = S^T f_{\text{ext}}(t), \quad \text{with} \quad M_q = S^T M S, \quad K_q = S^T K S. \qquad (4.2)$$

Due to orthogonality, the matrices $M_q = \text{diag}\{m_n\}_{n=1}^N$ and $K_q = \text{diag}\{k_n\}_{i=n}^N$ are diagonal. Therefore the *modal displacements* $\{q_n\}_{n=1}^N$ obey a second-order linear oscillator equation with frequencies $\omega_n^2 = k_n/m_n$, where m_n and k_n represent the modal masses and stiffnesses, and where the transposed matrix S^T defines how a driving force f_{ext} acts on the modes. The oscillation $y_l(t)$ at the lth spatial point is the sum of the modal oscillations weighed by the modal shapes: $y_l(t) = \sum_{n=1}^N s_{n,l} q_n(t)$.

Equivalently, modal decomposition can be obtained from the partial differential equation that describes a distributed object, in which the displacement $y(x, t)$ is a continuous function of space and time. In this case, a normal mode is a factorized solution $y(x, t) = s(x)q(t)$. As an example, for a string with length L and fixed ends, the D'Alembert equation with fixed boundary conditions admits the factorized solutions $y_n(x, t) = s_n(x)q_n(t) = \sqrt{2/L}\sin(\omega_n t + \phi_n)\sin(k_n x)$, with $k_n = \frac{n\pi}{L}$ and $\omega_n = ck_n$ (c is the wave speed). If a force density $f_{\text{ext}}(x, t)$ is acting on the string, the equation is

$$\mu\frac{\partial^2 y}{\partial t^2}(x,t) - T\frac{\partial^2 y}{\partial x^2}(x,t) = f_{ext}(x,t), \qquad (4.3)$$

where T, μ are the string tension and density, respectively. Substituting the factorized solutions $y_n(x,t)$ and integrating over the string length yields

$$\left[\mu\int_0^L s_n^2(x)dx\right]\ddot{q}_n(t) + \left[T\int_0^L [s_n'(x)]^2 dx\right]q_n(t) = \int_0^L s_n(x)f_{ext}(x,t)dx. \qquad (4.4)$$

Therefore the equation for the nth mode is that of a second-order oscillator with mass $m_n = \mu\int_0^L s_n^2(x)dx$ and stiffness $k_n = T\int_0^L [s_n'(x)]^2 dx$. The modal shape defines how the external force acts on the mode, and the oscillation $y(x_{out},t)$ of the system at a given spatial point x_{out} is the sum of the modal oscillations weighed by the modal shapes: $y(x_{out},t) = \sum_{n=1}^{+\infty} s_n(x_{out})q_n(t)$.

The two (discrete and continuous) modal representations of oscillating systems have strict analogies, reflecting the fact that continuous systems can be seen as the limit of discrete systems when the number of masses becomes infinite. As an example, a string can be approximated with the discrete network of Fig. 4.5, with N masses. The discrete system has N modes, whose shapes resemble more and more closely those of the continuous system, as N increases.

The modal formalism can be extended to systems that include damping, i.e. where a term $R\dot{y}$ is added in Eq. 4.1, or the terms $d_1\partial y/\partial t + d_2\partial/\partial t(\partial^2 y/\partial x^2)$ are added on the left-hand side of Eq. 4.3. However, certain hypotheses about the damping matrix must hold.

Given the modal decomposition for a certain resonating object, sound synthesis can be obtained from a parallel structure of second-order numerical oscillators, each representing a particular mode. Despite the comparatively low computational costs with respect to other techniques, mode-based numerical schemes can become expensive when many objects, each with many modes, are impacted simultaneously. Therefore recent studies deal with optimization of modal synthesis schemes. Bonneel et al. (2008) proposed an approach based on short-time Fourier Transform, that exploits the inherent sparsity of modal sounds in the frequency domain. Other

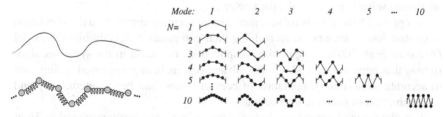

Fig. 4.5 Analogies between continuous and discrete systems. *Left* Approximation of an ideal string with a mass-spring network. *Right* Modes of the discrete system for various numbers N of masses

research has used perceptual criteria to perform mode compression and truncation, so as to reduce the computational load (Raghuvanshi and Lin 2007; Lloyd et al. 2011). The use of variable time-steps in the integration of the numerical equation has also been explored (Zheng and James 2011).

Another current area of research concerns improvements to the basic modal synthesis scheme, with the aim of increasing the realism and quality of the resulting sounds. One of the key challenges is the development of automatic modal analysis and determination of material parameters that recreate realistic audio. Ren et al. (2013) presented a method that analyses pre-recorded target audio clips to estimate perceptually salient modal parameters that capture the inherent quality of recorded sounding materials. A different approach was proposed by Picard et al. (2010), in which automatic voxelization of a surface model is performed, and automatic tuning of the corresponding finite element method parameters is obtained based on the distribution of material in each cell.

4.3.2 Impact Forces

If the external force applied to the resonating object is an ideal impulse, the oscillation is a weighed sum of damped sinusoids. More in general, energy is injected into the system through some kind of excitation mechanism. The amount and the rate at which energy enters the system depends on the nature of the interaction. Impact is a relatively simple interaction, as it occurs in a quasi-impulsive manner, rather than entailing continuous exchange of energy (as it happens for rolling, scraping, stick-slip friction, and so on). At the simplest level, a feed-forward scheme can be used in which the resonator is set into oscillation by driving forces that are externally computed or recorded. As an example, the contact force describing an impact onto a resonating object may be modeled with the following signal (van den Doel and Pai 2004):

$$f(t) = \begin{cases} \frac{f_{max}}{2}\left[1 - \cos(\frac{2\pi t}{\tau})\right], & 0 \le t \le \tau, \\ 0, & t > \tau. \end{cases} \tag{4.5}$$

Here, the time-dependent force signal has a cosinusoidal shape in which the duration of the force (i.e., the contact time) is determined by the parameter τ, while its maximum value is set using the parameter f_{max}.

As opposed to feed-forward schemes, a more accurate approach to the simulation of contact forces amounts to embedding their computation directly into the model (Avanzini et al. 2003). Despite the complications that arise in the synthesis algorithms, this approach provides some advantages, including improved quality due to accurate audio-rate computation of contact forces, and better interactivity and responsiveness of sound to user actions.

A model for the impact force between two objects, originally proposed by Hunt and Crossley (1975), is the following:

Fig. 4.6 Non-linear
force (4.6) generated during impact of a point mass
on a hard surface, for various
impact velocities

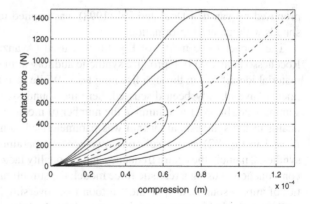

$$f(x(t), v(t)) = \begin{cases} kx(t)^\alpha + \lambda x(t)^\alpha \cdot v(t) & x > 0, \\ 0 & x \leq 0, \end{cases} \quad (4.6)$$

where the time-dependent function $x(t)$ is the interpenetration of the two colliding
objects (representing the overall surface deformation of the two objects during contact), and $v = \dot{x}$ is the corresponding velocity. In this case the dynamics of the contact
force is not pre-determined as in Eq. 4.5, but is itself dependent on the states and the
oscillations of the objects. The parameters k, λ control the stiffness of the impact and
the involved dissipation, while the exponent α is related to the local geometry around
contact [in particular, $\alpha = 3/2$ in the classic Hertz model of collision between two
spheres (Flores et al. 2008)]. Figure 4.6 depicts the simulation of a point mass hitting
a rigid surface with the impact model of Eq. 4.6: it can be noted that the impact
force has a non-linear characteristics that depends on the exponent α, and exhibits a
hysteresis effect that is associated to the dissipative component.

Several refinements to these models have been proposed in order to improve the
sound quality. Other relevant phenomena occurring during the impact need to be simulated, particularly the acceleration noise produced as a consequence of large rigid-
body accelerations over a short time scale, which results in a perceivable acoustic
pressure disturbance at the attack transient (Chadwick et al. 2012).

4.3.3 Rendering of Materials and Hardness in Impact Sounds

As discussed previously, material properties of objects can be perceived auditorily
from impact sounds. In particular, object hardness/softness correlates strongly with
material identification and impact stiffness.

The modal representation of a resonating object is naturally linked to many ecological dimensions of the corresponding sounds. The possibility of linking the physical
model parameters to sound parameters related to the perception of material was
first demonstrated by Klatzky et al. (2000). In this work, the modal representation

proposed by van den Doel and Pai (1998) was applied to the synthesis of impact sounds with material information.

The impact force model of Eq. 4.6 was used (Avanzini and Rocchesso 2001; Rocchesso et al. 2003) to produce synthetic auditory stimuli for the investigation of material identification through listening tests. While results from perceptual experiments have already been discussed, here the sound synthesis approach is briefly summarized. In order to minimize the number of model parameters, the modal resonator in the synthesis algorithm was parameterised to have only one mode (i.e., $N = 1$ in Eq. 4.2): as a result, only two acoustic parameters controlled the sound synthesis, namely the centre frequency and the quality factor of the single mode. As a consequence of using a realistic force model, the stimuli maintained the characteristics of impact sounds, despite the "cartoon-like" oversimplification of the resonator.

Ren et al. (2013) investigated the use of the Rayleigh damping model in modal sound synthesis. As discussed in the previous section, modal decoupling is only feasible under certain assumptions for the damping matrix. Rayleigh damping is a commonly adopted approximation model that enables such decoupling, and assumes the damping matrix to be a linear combination of the mass and stiffness matrices. With the goal of investigating whether auditory material perception under the Rayleigh damping assumption is geometry-invariant (i.e., whether this approximation is transferable across different shapes and sizes), Ren et al. (2013) used both real-world audio recordings and (modal) synthesized sounds to perform objective and subjective analysis of the validity of the Rayleigh damping model across different object shapes and sizes. Statistical analysis showed that this is the case for homogeneous materials, whereas the Rayleigh damping model does not provide equally good approximation for materials with heterogeneous micro-structures, such as wood. This study therefore points to some often overlooked limitations of modal sound synthesis.

Aramaki et al. (2011) proposed a modal-based synthesizer of impact sounds, controlled through high-level verbal descriptors referring to material categories (i.e., wood, metal and glass). Analysis was conducted on a set of acoustical descriptors (namely, attack time, spectral bandwidth, roughness, and normalized sound decay), together with electrophysiological measurements (in particular, analysis of changes in brain electrical activity using event related potentials). Based on acoustic and electrophysiological results, a three-layer control architecture providing the synthesis of impact sounds directly from the material label was proposed.

In a study on material perception in a bimodal virtual environment, specifically on the mutual interaction of audio and graphics, Bonneel et al. (2010) also used the modal approach proposed by van den Doel et al. (2001). An experiment similar to the one reported by Klatzky et al. (2000) was conducted. Results showed that the proposed bimodal rendering resulted in better perceived quality for both audio and graphics, and that there was a mutual influence of the two modalities on the perception of material similarity.

Recent research has addressed the issue of auditory rendering of materials from the point of view of walking interactions (Steinicke et al. 2013). Fontana and co-workers (Visell et al. 2009; Marchal et al. 2013) published many results about footstep sound design, including the rendering of floor surface material, as well as multimodal issues and the integration of vibrotactile display. All sounds were designed using the Sound Design Toolkit (SDT), a software package providing a palette of virtual lutheries and foley pits, targeted at sonic interaction design research and education (Delle Monache et al. 2010). Real-time synthesis of footstep sounds for various materials was also investigated by Nordahl et al. (2010), using modal synthesis and the impact model of Eq. 4.6. A perceptual study was conducted with three groups of subjects: the first group listened to recorded footstep sounds, the second one generated synthetic footstep sounds interactively by walking on shoes augmented with sensors, and the third one listened to pre-recorded footstep sounds synthesized with the same synthesis engine. Results showed that subjects of the second group were able to identify synthesized floor materials at a comparable accuracy with real-world recordings, while the performance of the third group was significantly worse.

Overall, the studies reviewed up to this point show that modal sound synthesis is the most commonly used approach in current literature to render auditorily different materials in impact sounds. There is a trend toward the definition of higher-level control parameters, that refer to ecological categories and hide low-level modal parameters in the background (Aramaki et al. 2011; Delle Monache et al. 2010). Some limitations of this approach have also been highlighted (Ren et al. 2013).

The proportion of studies devoted to the auditory rendering of object hardness/softness is comparatively low with respect to those devoted to material identification. In a modal synthesizer, contact hardness should be rendered by properly adjusting the parameters of the contact force in order to control the hardness-related auditory parameters discussed in the previous sections.

If a physical model such as the one in Eq. 4.6 is used to describe the contact force, an analytical expression for the contact time can be derived (Avanzini and Rocchesso 2004; Papetti et al. 2011):

$$\tau = \left(\frac{m}{k}\right)^{\frac{1}{\alpha+1}} \cdot \left(\frac{\mu^2}{\alpha+1}\right)^{\frac{\alpha}{\alpha+1}} \cdot \int_{v_{out}}^{v_{in}} \frac{1}{(1+\mu v)\left[-\mu(v-v_{in}) + \log\left|\frac{1+\mu v}{1+\mu v_{in}}\right|\right]^{\frac{\alpha}{\alpha+1}}},$$
(4.7)

where m is the hammer mass, while the remaining parameters are part of Eq. 4.6. This equation states that the contact time τ depends only on μ, the exponent α and the ratio m/k, in addition to the impact velocity v_{in}. Since neither m nor k affect the value of the integral, it follows that, for a given value of v_{in}, the dependence $\tau \sim (m/k)^{1/(\alpha+1)}$ holds.

Based on this analytical property, a relation between the contact time and the time-varying spectral centroid of the impact sound was discussed by Avanzini and Rocchesso (2004). As a result, a mapping between the physical parameters of the impact force and the hardness-related auditory parameters was proposed. Avanzini and Crosato (2006a) tested this relation in a bimodal (audio-haptic) setting. A subjective test was conducted in which subjects had to tap on virtual audio-haptic surfaces. In each condition the haptic stiffness had the same value while the acoustic stiffness k was varied. Perceived hardness/softness was determined using an absolute magnitude-estimation procedure. Results showed that subjects consistently ranked the surfaces according to the values of k in the auditory stimuli. If the impact force is not described physically but rather with a signal model, similar control can be achieved. As an example, the cosinusoidal impact force model in Eq. 4.5 includes the contact time τ among its parameters.

As a conclusion to this section, it may be argued that other features of an impact model should be adjusted to properly render the perception of object hardness/softness. As an example, the amount of acceleration noise at the attack (Chadwick et al. 2012) is related to the contact stiffness. Moreover, more complex impact force models (e.g. distributed models taking into account the contact area) may be needed for a more accurate rendering of the attack transient.

4.3.4 Rendering of Deformable and Aggregate Objects

It should be no surprise that the particular topic of rendering sounds produced by interaction with deformable objects, like textiles, tissues, and so on, has seen very little work. In fact, as already noted previously in this chapter, softness-related auditory information is less accessible when interacting with soft objects than with hard ones.

A few studies deal with the topic of textile sounds synthesis, although the relation between the sound rendering and the perceived softness is not investigated. Existing studies have an applicative focus, in which multimodal rendering of textiles is used for high-quality animation and possibly for enhanced active exploration of virtual fabrics (e.g. for e-commerce applications).

Huang et al. (2003) proposed an audio-haptic interface for simulating interaction with a fabric surface through a stylus. The exploratory procedure considered in this work was mainly rubbing of the stylus over a cloth patch. Sound was synthesized using a modal model driven by measured roughness profiles. While appropriate for the particular haptic application considered in this work, the model is hardly generalizable to more complex cloth animations.

It has already been mentioned previously that sliding friction due to textile rubbing against itself or other surfaces is an important component of textile sound but is not the only one. Moreover, frictional sounds are related to surface geometry properties of the object, rather than material properties. In addition to frictional sounds, textiles can also buckle and produce crumpling sounds, in the form of small audible pops. Woven garments produce audible crumpling sounds, while stiff synthetic clothes (e.g. nylon windbreakers), exhibit characteristically loud crumpling sounds. An et al. (2012) proposed a data-driven method for automatically synthesizing sound in physics-based cloth animations. Given a cloth animation, analysis of the deformation was used to drive crumpling and friction sound models estimated from cloth measurements and to synthesize low-quality audio. This was then used as a target signal for a sound synthesis process, which selected best-match short segments from a database of recorded cloth sounds.

Crumpling sounds, such as those used by An et al. (2012) as a component of cloth sound synthesis, are another interesting category of sounds related to softness/hardness perception. An example are the sounds produced by crumpling paper, which can be modeled in terms of (i) the probabilistic distribution of the energies of the short transients, and (ii) a model of the temporal density of transients as a stationary Poisson process (Houle and Sethna 1996). This approach has inspired the development of geometry-independent stochastic models of crumpling, which were used to design sounds produced by deformations of aggregate materials, such as sand, snow, or gravel (Fontana and Bresin 2003). Such sounds belie a common temporal process originating with the transition toward a minimum-energy configuration of an ensemble of microscopic systems, by way of a sequence of transient events. Models of this type have been used in particular to mimic the sound of a footstep onto aggregate grounds (Fontana and Bresin 2003; Marchal et al. 2013; Nordahl et al. 2010; Visell et al. 2009).

This brief overview shows that there is wide space for novel research on the synthesis of sounds of aggregate and deformable materials. The remainder of this section reviews a few studies in which auditory information related to object softness is rendered using different, non-ecological forms of auditory feedback, mostly through some kind of sonification of haptic signals. It may be argued that the adoption of such non-ecological approaches is due to the scarce availability and exploitability of auditory information in the interaction with very soft materials.

Yao et al. (2005) developed a probe to enhance tactile sensations experienced during surgery, specifically during tissue examination with minimally invasive procedures. The probe detects and magnifies the acceleration signal resulting from the interaction of its tool tip with the tissue surfaces. Since the acceleration signal is highly structured and spectrally rich, auditory feedback was obtained through direct conversion of this signal into audio. In the literature of auditory display, this particular approach to sonification is known as "audification" (Dombois and Eckel 2011). Subjective experiments under various conditions (with no amplification, with enhanced tactile feedback, with sound feedback, and with passive touch) showed significant improvements in the recognition of tissue features in the case of tactile and auditory feedbacks.

Kitagawa et al. (2005) performed subjective tests on the sensory substitution of force feedback with sound, in the context of a robotic surgical system. The sound design is not explained in detail: the authors write about a "single tone" (possibly a sinusoid or other waveform) to be played back when the tension applied by the operator exceeded a target value. It was reported that this type of sensory substitution provided statistically significant improvements in applied force accuracy and consistency during the performance of a simple surgical task.

An attempt to formalize a unified approach to study the relationship between physical parameters and coding parameters used to convey control information through the auditory modality was provided by Csapo and Baranyi (2010). The proposed sonification formalism was demonstrated through an application in which the physical properties of a surface are conveyed to a remote teleoperator through sound. Softness/hardness properties, in particular, were continuously sonified using frequency-modulated or amplitude-modulated signals. However, no subjective tests were conducted to assess the effectiveness of the proposed sonification.

4.3.5 Rendering of Liquid Sounds

Compared to the amount of studies devoted to the sound synthesis and rendering of mechanical interactions between solid objects, those dealing with liquid sounds are a small proportion.

Given the great variety of possible liquid sounds (ranging from stochastic sounds such as that of streaming river, to deterministic ones such as dripping), their synthesis remains a complicated task. Existing research has focused on simulating some of the specific mechanisms responsible for sound generation in liquids, particularly bubble formation. After being formed in a liquid, a bubble emits a decaying sinusoidal sound. If bubble formation occurs close enough to the liquid-air interface, the pitch rises as it approaches the surface. The physical mechanism responsible for these sounds is the pulsation of the bubble volume (Minnaert 1933): any bubble being a small compressible air region surrounded by incompressible fluid, it oscillates like a spring amid a liquid domain.

A few recent studies have dealt with bubble sound synthesis. The first model was proposed in van den Doel's seminal work (van den Doel 2004, 2005). Starting from the physical description provided by Minnaert (1933), a simple algorithm was developed to synthesize single bubbles, using such physical control parameters as bubble radius, loss coefficient, and velocity. Being the model extremely efficient, a real-time bubble simulator was realized, which allowed simulation of more complex liquid sounds (from dripping to heavy rain or waterfalls) through synthesis of a large population of bubbles. The realism of the model was preliminary tested with subjects in a listening experiment. Results suggested that bubbles with radii in the range $2 - 7$ mm are most readily associated with the sound of a water drop, and that the rising pitch increases the realism of larger (>4 mm) rather than smaller bubbles, consistently with the fact that these have very high pitch and decay very rapidly. Very large bubbles sounded unnatural, consistently with the fact that they do not occur in isolation in nature.

Zheng and James (2009) proposed a similar approach to acoustic bubble simulation, with the aim of augmenting existing numerical solvers for incompressible liquid simulations that are commonly adopted in the computer graphics literature. The proposed model included bubble advection, time-dependent pitch, and a simplified description of the bubble entrainment process. Sound radiation was modeled through a time-varying sum of bubble oscillators, weighted by their acoustic transfer function modeled as a discrete Green's function of the Helmholtz equation. A fast numerical solver was proposed, which allowed simulation of large numbers of bubbles. Examples for various liquid sounds were proposed (including pouring, babbling, and splashing phenomena), although no psychophysical validation was presented.

Moss et al. (2010) also proposed a simplified, physically inspired model for bubble creation, designed specifically for real-time applications. The model used the fluid surface curvature and velocity as parameters for bubble creation and a stochastic model for bubble sound synthesis based on van den Doel's work (van den Doel 2004, 2005). A user study was conducted to assess the realism of various types of liquid sounds synthesized using the proposed approach. Results suggested that the perceived realism is comparable to recorded sounds in similar settings. However, the model was designed for a shallow water simulator, which reduces interaction possibilities by allowing only surface waves, precluding splashes and object penetration.

In order to bridge the complexity of fluid-dynamic simulations with the needs of interactive sonification, Drioli and Rocchesso (2012) proposed a multi-rate approach to the sound synthesis of liquid phenomena, in which smoothed particle motion simulated at low-rate is used to model liquids in motion and to control audio-rate sound synthesis algorithms of basic acoustic events. Two such basic events were simulated, namely bubbles and surface impacts. In this way, a larger family of sounds can be rendered, including liquid-liquid and liquid-solid interactions. The approach was illustrated through two configurations: the falling of a liquid volume into a container, and the falling of a solid object into a container filled with liquid at rest.

More recently, Cirio et al. (2013) introduced the use of vibrotactile feedback as a rendering modality for solid-fluid interaction, based on the associated sound generating physical processes. Similarly to earlier works, sound was generated from bubble simulation inspired by Moss et al. (2010) and based on a particle-based fluid model (Monaghan 1992). A novel vibrotactile model was then introduced, which received events from the physical simulation and synthesized a signal through three different components: a high-frequency component produced by initial impact of an object onto the liquid, components due to oscillations of smaller bubbles, and the main cavity oscillation. A pilot study was conducted to assess the perceived interaction qualitatively.

As a conclusion to this section, it should be noted that, apart from preliminary user tests aimed at assessing the perceived realism of the proposed simulations, none of the above studies included more extensive psychophysical experiments on the ability of listeners to estimate specific properties of synthesized liquid sound sources.

4.4 Conclusion

Material properties have a front-row seat in the theoretical and empirical study of non-vocal everyday sound sources. Source-perception research has revealed a great deal about the strengths and weaknesses of the auditory estimation of material properties, about how material properties interact perceptually with other mechanical properties of sound-generating events, and about the acoustical factors that underlie perceptual judgments. Most of the research up to this point has focused on stiff solid objects, and has largely disregarded deformable materials such as fabrics or liquids. For both of these, it is thus still unclear the extent to which source-perception processes might actually rely on material properties rather than material-independent properties such as the texture-defining geometry of fabrics, or the temporally variable geometry of sound-generating bubbles in a liquid. The study of auditory materials has witnessed a number of interesting recent developments that show promising potentials for future research. Research on auditory contexts has begun to unravel the factors involved in the more general ability to differentiate between states of matter of sound-generating substances, and to benchmark the perception of materials against that of the properties of sound-generating interactions. Research on audio-haptic contexts has begun to address the interactions between material information presented in different modalities from both the perceptual and motor-control points of view. Further, promising directions of research are also represented by the study of the cortical processes involved in the processing of material-related information (Arnott et al. 2008; Aramaki et al. 2010; Micoulaud-Franchi et al. 2011).

Studies in ecological acoustic have been re-discovered in the late 1990's in the light of sound design and sound rendering for virtual reality, and have been a major driver for research on the synthesis non-vocal everyday sounds. Techniques for modal synthesis of sounds produced by stiff objects in impulsive or continuous contact are now well established. It can be expected that upcoming research will continue to focus on the development of more refined and realistic models of the interaction. Due to the impulsive and highly non-linear nature of impact forces, one current open issue concerns the definition of specialized numerical techniques for the accurate simulation of such forces (Papetti et al. 2011; Chatziioannou and van Walstijn 2013). Further improvements in realism will be achieved through the simulation of secondary physical mechanisms involved in the interaction, such as acceleration noise and its relation to the contact stiffness (Chadwick et al. 2012), as well as the effects of distributed and possibly time-varying contact areas.

In the mid to long term, it can be expected that other physical modeling techniques, such as time-domain finite differences (FDTD) and finite element methods (FEM) will gain popularity and become competitive with modal synthesis. Being "brute force" approaches, they possess the advantage of generality since a great variety of systems can be approached without the need for making simplifying hypotheses or adding intermediate levels of representation. Material properties in particular are completely controllable, since all low-level material-related parameters (Young and shear modulus, Poisson coefficient, density, etc.) are directly embedded into the models. On the other hand, such methods are numerically intensive: future research will therefore be devoted to looking at efficient implementations particularly in parallel architectures (multicore processors and general purpose graphics processing units, see e.g. Bilbao et al. 2013). Improvements in sound quality promise to be striking, however only very recently has computational power grown to the extent that sound can be synthesized in a reasonable amount of time with these techniques, and real-time is still a long way off.

While impacts between stiff objects have been thoroughly studied, there is wide space for novel research on the synthesis of other categories of sounds. Our review has shown that there is a handful of research on the synthesis of sounds produced by deformable objects. In particular, the most recent studies on textile and cloth sounds illustrate the many issues involved in the synthesis of such complex sounds, and at the same time demonstrate the potential for research in this direction. Similar considerations apply to liquid sound synthesis, although in this case the number of existing studies is marginally larger. In both cases, there is a lack of validation of the proposed approaches in terms of their ability to convey specific sound source properties to the listener.

Acknowledgments This work was partially supported by the Marie Curie Intra-European Fellowships program (FP7 PEOPLE-2011-IEF-30153, project BrainInNaturalSound to Bruno L. Giordano). The authors wish to thank Laurie Heller and Guillaume Lemaitre for sharing the sound stimuli used to prepare Figs. 4.1 and 4.3, and Laurie Heller, Federico Fontana and Stephen McAdams for providing helpful feedback about earlier versions of this chapter.

References

Adrien J-M (1991) The missing link: modal synthesis. In: De Poli G, Piccialli A, Roads C (eds) Representations of musical signals. MIT Press, Cambridge, pp 269–297

An SS, James DL, Marschner S (2012) Motion-driven concatenative synthesis of cloth sounds. ACM Trans Graph (TOG) 31(4) (Article No. 102)

Aramaki M, Marie C, Kronland-Martinet R, Ystad S, Besson M (2010) Sound categorization and conceptual priming for nonlinguistic and linguistic sounds. J Cognit Neurosci 22:2555–2569

Aramaki M, Besson M, Kronland-Martinet R, Ystad S (2011) Controlling the perceived material in an impact sound synthesizer. IEEE Trans Audio Speech Lang Process 19(2):301–314

Arnott SR, Cant JS, Dutton GN, Goodale MA (2008) Crinkling and crumpling: an auditory fmri study of material properties. Neuroimage 43:368–378

Ashby FG (1992) Multidimensional models of perception and cognition. Lawrence Erlbaum Associates, Hills-dale

Attias H, Schreiner CE (1997) Temporal low-order statistics of natural sounds. In: Mozer MC, Jordan MI, Petsche T (eds) Advances in neural information processing systems 9. MIT Press, Cambridge, pp 27–33

Avanzini F, Rath M, Rocchesso D, Ottaviani L (2003) Low-level sound models: resonators, interactions, surface textures. In: Rocchesso D, Fontana F (eds) The sounding object. Mondo Estremo, Firenze, pp 137–172

Avanzini F, Crosato P (2006b) Integrating physically-based sound models in a multimodal rendering architecture. Comput Anim Virtual Worlds 17(3–4):411–419. doi:10.1002/cav.v17:3/4

Avanzini F, Crosato P (2006a) Haptic-auditory rendering and perception of contact stiffness. In: McGookin D, Brewster S (eds) Haptic and audio interaction design. Lecture Notes in Computer Science 4129/2006, Springer, Berlin. pp 24–35

Avanzini F, Rocchesso D (2001) Controlling material properties in physical models of sounding objects. In: Proceedings of the international computer music conference (ICMC'01). La Habana, pp 91–94 (Available at http://www.soundobject.org)

Avanzini F, Rocchesso D (2004) Physical modeling of impacts: theory and experiments on contact time and spectral centroid. In: Proceedings of the sound and music computing conference (SMC2004), Paris, pp 287–293

Ballas JA (1993) Common factors in the identification of an assortment of brief everyday sounds. J Exp Psychol Hum Percept Perform 19:250–267

Belin P, Zatorre RJ, Lafaille P, Ahad P, Pike B (2000) Voice-selective areas in human auditory cortex. Nature 403:309–312

Bilbao S (2009) Numerical sound synthesis—finite difference schemes and simulation in musical acoustics. Wiley, Chichester

Bilbao S, Hamilton B, Torin A, Webb C, Graham P, Gray A, Perry J (2013) Large scale physical modeling sound synthesis. In: Proceedings of the Stockholm music acoustic conference (SMAC2013), Stockholm, pp 593–600

Bonneel N, Drettakis G, Tsingos N, Viaud-Delmon I, James D (2008) Fast modal sounds with scalable frequency-domain synthesis. ACM Trans Graph (TOG) 27(3) (Article no. 24)

Bonneel N, Suied C, Viaud-Delmon I, Drettakis G (2010) Bimodal perception of audio-visual material properties for virtual environments. ACM Trans Appl Percept 7(1) (Article No. 1)

Borg I, Groenen P (1997) Modern multidimensional scaling. Springer, New York

Cabe PA, Pittenger JB (2000) Human sensitivity to acoustic information from vessel filling. J Exp Psychol Hum Percept Perform 26:313–324

Carello C, Wagman JB, Turvey MT (2003) Acoustical specification of object properties. In: Anderson J, Anderson B (eds) Moving image theory: ecological considerations. Southern Illinois University Press, Carbondale

Castiello U, Giordano BL, Begliomini C, Ansuini C, Grassi M (2010) When ears drive hands: the influence of contact sound on reaching to grasp. PLoS ONE 5:e12240

Chadwick JN, Zheng C, James DL (2012) Precomputed acceleration noise for improved rigid-body sound. ACM Trans Graph (TOG)31(4) (Article No. 103)

Chatziioannou V, van Walstijn M (2013) An energy conserving finite difference scheme for simulation of collisions. In: Proceedings of the sound and music computing conference (SMC2013), Stockholm, pp 584–591

Cho G, Casali JG, Yi E (2001) December). Effect of fabric sound and touch on human subjective sensation. Fibers Polym 2(4):196–202

Cho G, Kim C, Cho J, Ha J (2005) March). Physiological signal analyses of frictional sound by structural parameters of warp knitted fabrics. Fibers Polym 6(1):89–94

Cirio G, Marchal M, Lécuyer A, Cooperstock J (2013) Vibrotactile rendering of splashing fluids. ACM Trans Haptics 6(1):117–122

Civille GV, Dus CA (1990) Development of terminology to describe the handfeel properties of paper and fabrics. J Sen Stud 5(1):19–32

Csapo AB, Baranyi P (2010) An interaction-based model for auditory substitution of tactile percepts. In: Proceedings of IEEE international conference on intelligent engineering systems (INES 2010), Las Palmas of Gran Canaria, pp 248–253

De Coensel B, Vanwetswinkel S, Botteldooren D (2011) Effects of natural sounds on the perception of road traffic noise. J Acoust Soc Am 129:EL148–EL153

Delle Monache, S., Polotti, P., & Rocchesso, D. (2010, September). A toolkit for explorations in sonic interaction design. In: Proceedings of audio mostly conference (AM'10). Piteå (Article no. 1)

DiFranco DE, Beauregard GL, Srinivasan MA (1997) The effect of auditory cues on the haptic perception of stiffness in virtual environments. In: Proceedings of the ASME dynamic systems and control division, DSC, vol 61, pp 17–22

Dombois F, Eckel G (2011) Audification. In: Hermann T, Hunt A, Neuhoff JG (eds) The sonification handbook. Logos Verlag, Berlin, pp 301–324

Doutaut V, Matignon D, Chaigne A (1998) Numerical simulations of xylophones. II. Time-domain modeling of the resonator and of the radiated sound pressure. J Acoust Soc Am 104(3):1633–1647

Drioli C, Rocchesso D (2012) Acoustic rendering of particle-based simulation of liquids in motion. J Multimodal User Interfaces 5(3–4):187–195

Ernst MO, Banks MS (2002) Humans integrate visual and haptic information in a statistically optimal fashion. Nature 415:429–433

Fletcher NH, Rossing TD (1991) The physics of musical instruments. Springer, New York

Flores P, Claro JP, Lankarani HM (2008) Kinematics and dynamics of multibody systems with imperfect joints: Models and case studies. Springer, Berlin

Fontana F, Bresin R (2003) Physics-based sound synthesis and control: crushing, walking and running by crumpling sounds. In: Proceedings of the colloquium on music informatics (CIM 2003), Firenze, pp 109–114

Franinović K, Serafin S (eds) (2013) Sonic interaction design. MIT Press, Cambridge

Freed DJ (1990) Auditory correlates of perceived mallet hardness for a set of recorded percussive events. J Acoust Soc Am 87:311–322

Gaver WW (1988) Everyday listening and auditory icons. Unpublished doctoral dissertation, University of California, San Diego

Gaver WW (1993) What in the world do we hear? an ecological approach to auditory event perception. Ecol Psychol 5:1–29

Geffen MN, Gervain J, Werker JF, Magnasco MO (2011) Auditory perception of self-similarity in water sounds. Front Integr Neurosci 5

Gibson JJ (1966) The senses considered as a perceptual system. Houghton Mifflin, Boston

Gibson JJ (1979) The ecological approach to visual perception. Houghton Mifflin, Boston

Giordano BL, Avanzini F, Wanderley M, McAdams S (2010) Multisensory integration in percussion performance. In: Actes du 10eme congrès français d'acoustique, Lyon (p. [CD-ROM]). Société Française d'Acoustique, Paris, France

Giordano BL (2005) Sound source perception in impact sounds. Unpublished doctoral dissertation, University of Padova, Italy

Giordano BL (2003) Material categorization and hardness scaling in real and synthetic impact sounds. In: Rocchesso D, Fontana F (eds) The sounding object. Mondo Estremo, Firenze, pp 73–93

Giordano BL, McAdams S (2006) Material identification of real impact sounds: effects of size variation in steel, glass, wood and plexiglass plates. J Acoust Soc Am 119:1171–1181

Giordano BL, McDonnell J, McAdams S (2010) Hearing living symbols and nonliving icons: category-specificities in the cognitive processing of environmental sounds. Brain Cogn 73:7–19

Giordano BL, Rocchesso D, McAdams S (2010) Integration of acoustical information in the perception of impacted sound sources: the role of information accuracy and exploitability. J Exp Psychol Hum Percept Perform 36:462–479

Giordano BL, Guastavino C, Murphy E, Ogg M, Smith BK, McAdams S (2011) Comparison of methods for collecting and modeling dissimilarity data: applications to complex sound stimuli. Multivar Behav Res 46:779–811

Giordano BL, Visell Y, Yao HY, Hayward V, Cooperstock J, McAdams S (2012) Identification of walked-upon materials in auditory, kinesthetic, haptic and audio-haptic conditions. J Acoust Soc Am 131:4002–4012

Grassi M (2005) Do we hear size or sound? Balls dropped on plates. Percept Psychophys 67:274–284

Gygi B, Kidd GR, Watson CS (2007) Similarity and categorization of environmental sounds. Percept Psychophys 69:839–855

Handel S, Erickson ML (2001) A rule of thumb: the bandwidth for timbre invariance is one octave. Music Percept 19:121–126

Houix O (2003) Categorisation auditive des sources sonores. Unpublished doctoral dissertation, Université du Maine, France

Houix H, Lemaitre G, Misdariis N, Susini P, Urdapilleta I (2012) A lexical analysis of environmental sound categories. J Exp Psychol Appl 18:52

Houle PA, Sethna JP (1996) Acoustic emission from crumpling paper. Phys Rev E 54(1):278–283

Huang G, Metaxas D, Govindaraj M (2003) Feel the "fabric": an audio-haptic interface. In: Proceedings of ACM siggraph/eurographics symposium on computer animation (SCA03). San Diego, pp 52–61

Hunt KH, Crossley FRE (1975) Coefficient of restitution interpreted as damping in vibroimpact. J Appl Mech 42:440–445

Jansson G (1993) Perception of the amount of fluid in a vessel shaken by hand. In: Valenti S, Pittinger J (eds) Studies in perception and action II. Posters presented at the VIIth international conference on event perception and action), pp 263–267

Jansson G, Juslin P, Poom L (2006) Liquid-specific stimulus properties can be used for haptic perception of the amount of liquid in a vessel put in motion. Perception 35:1421–1432

Jeon JY, Lee JL, You J, Kang J (2012) Acoustical characteristics of water sounds for soundscape enhancement in urban open spaces. J Acoust Soc Am 131:2101–2109

Kawabata S (1980) The standardization and analysis of hand evaluation. Textile Machinery Society of Japan, Osaka

Kidd GR, Watson CS (2003) The perceptual dimensionality of environmental sounds. Noise Control Eng J 51:216–231

Kitagawa M, Dokko D, Okamura AM, Yuh DD (2005) Effect of sensory substitution on suture-manipulation forces for robotic surgical systems. J Thorac Cardiovasc Surg 129(1):151–158

Klatzky RL, Pai DK, Krotkov EP (2000) Perception of material from contact sounds. Presence Teleoperators Virtual Environ 9(4):399–410

Kunkler-Peck AJ, Turvey MT (2000) Hearing shape. J Exp Psychol Hum Percept Perform 26:279–294

Lakatos S, McAdams S, Caussé R (1997) The representation of auditory source characteristics: simple geometric form. Percept Psychophys 59:1180–1190

Lambourg C, Chaigne A, Matignon D (2001) Time-domain simulation of damped impacted plates: II. Numerical Models and Results. J Acoust Soc Am 109(4):1433–1447

Lederman SJ (1979) Auditory texture perception. Perception 8(1):93–103

Lederman SJ, Klatzky RL (2004) Multisensory texture perception. In: Calvert G, Spence C, Stein B (eds) Handbook of multisensory processes. MIT Press, Cambridge, pp 107–122

Lemaitre G, Heller LM (2012) Auditory perception of material is fragile while action is strikingly robust. J Acoust Soc Am 131:1337–1348

Lemaitre G, Heller LM (2013) Evidence for a basic level in a taxonomy of everyday action sounds. Exp Brain Res 226(2):253–264

Lewis JW, Brefczynski JA, Phinney RE, Jannik JJ, DeYoe ED (2005) Distinct cortical pathways for processing tool versus animal sounds. J Neurosci 25:5148–5158

Lloyd B, Raghuvanshi N, Govindaraju NK (2011) Sound synthesis for impact sounds in video games. In: Proceedings of the ACM symposium on interactive 3D graphics and games (ACM I3D), San Francisco, pp 55–62

Lutfi RA, Oh EL (1997) Auditory discrimination of material changes in a struck-clamped bar. J Acoust Soc Am 102:3647–3656

Lutfi RA (2001) Auditory detection of hollowness. J Acoust Soc Am 110:1010–1019

Lutfi RA (2007) Human sound source identification. In: Yost WA, Fay RR, Popper AN (eds) Auditory Percept Sound Sources. Springer, New York, NY, pp 13–42

Lutfi RA, Liu CJ (2007) Individual differences in source identification from synthesized impact sounds. J Acoust Soc Am 122:1017–1028

Lutfi RA, Stoelinga CNJ (2010) Sensory constraints on auditory identification of the material and geometric properties of struck bars. J Acoust Soc Am 127:350–360

Marchal M, Cirio G, Visell Y, Fontana F, Serafin S, Cooperstock J, Lécuyer A (2013) Multimodal rendering of walking over virtual grounds. In: Steinicke F, Visell Y, Campos J, Lécuyer A (eds) Human walking in virtual environments. Springer, New York, pp 263–295

Marozeau J, de Cheveigné A, McAdams S, Winsberg S (2003) The dependency of timbre on fundamental frequency. J Acoust Soc Am 114:2946–2957

McAdams S, Chaigne A, Roussarie V (2004) The psychomecanics of simulated sound sources: material properties of impacted bars. J Acoust Soc Am 115:1306–1320

McAdams S, Roussarie V, Chaigne A, Giordano BL (2010) The psychomechanics of simulated sound sources: material properties of impacted thin plates. J Acoust Soc Am 128:1401–1413

Michaels CF, Carello C (1981) Direct perception. Prentice-Hall, Englewood Cliffs

Micoulaud-Franchi J-A, Aramaki M, Merer A, Cermolacce M, Ystad S, Kronland-Martinet R, Vion-Dury J (2011) Categorization and timbre perception of environmental sounds in schizophrenia. Psychiatry Res 189:149–152

Minnaert M (1933) On musical air-bubbles and the sounds of running water. Lond Edinb Dublin Philos Maga J Sci 16:235–248

Monaghan JJ (1992) Smoothed particle hydrodynamics. Ann Rev Astron Astrophys 30:543–574

Moss W, Yeh H, Hong J-M, Lin MC, Manocha D (2010) Sounding liquids: automatic sound synthesis from fluid simulation. ACM Trans Graph 29(3) (Article No. 21)

Nordahl R, Serafin S, Turchet L (2010) Sound synthesis and evaluation of interactive footsteps for virtual reality applications. In: Proceedings of the IEEE international conference on virtual reality (VR 2010), Waltham, pp 147–153

O'Brien JF, Shen C, Gatchalian CM (2002) Synthesizing sounds from rigid-body simulations. In: Proceedings of the 2002 ACM SIGGRAPH/Eurographics symposium on computer animation. San Antonio, TX, pp 175–181

Papetti S, Avanzini F, Rocchesso D (2011) Numerical methods for a non-linear impact model: a comparative study with closed-form corrections. IEEE Trans Audio Speech Lang Process 19(7):2146–2158

Picard C, Frisson C, Faure F, Drettakis G, Kry P (2010) Advances in modal analysis using a robust and multiscale method. EURASIP J Adv Sig Process 2010 (Article ID 392782.)

Raghuvanshi N, Lin MC (2007) Physically based sound synthesis for large-scale virtual environments. Comput Graph Appl IEEE 27(1):14–18

Ren Z, Yeh H, Klatzky R, Lin MC (2013) Auditory perception of geometry-invariant material properties. IEEE Trans Vis Comput Graph 19(4):557–566

Ren Z, Yeh H, Lin MC (2013) Example-guided physically based modal sound synthesis. ACM Trans Graph (TOG) 32(1) (Article No. 1)

Rocchesso D, Ottaviani L, Fontana F, Avanzini F (2003) Size, shape, and material properties of sound models. In: Rocchesso D, Fontana F (eds) The sounding object. Mondo Estremo, Firenze, pp 95–110

Rocchesso D (2011) Explorations in sonic interaction design. Logos Verlag, Berlin

Smith JO III (2004) Virtual acoustic musical instruments: review and update. J New Music Res 33(3):283–304

Sreng J, Bergez F, Legarrec J, Lécuyer A, Andriot C (2007) Using an event-based approach to improve the multimodal rendering of 6dof virtual contact. In: Proceedings of the ACM symposium on virtual reality software and technology (VRST07), Newport Beach, CA, pp 165–173

Steele KM, Williams AK (2006) Is the bandwidth for timbre invariance only one octave? Music Percept 23:215–220

Steinicke F, Visell Y, Campos J, Lecuyer A (eds) (2013) Human walking in virtual environments: perception, technology, and applications. Springer, New York

Tucker S, Brown GJ (2003) Modelling the auditory perception of size, shape and material: applications to the classification of transient sonar sounds. In: Proceedings of the 114th convention of the Audio Engineering Society

Turchet L, Serafin S (2014) Semantic congruence in audio-haptic simulation of footsteps. Appl Acoust 75:59–66

Turchet L, Serafin S, Cesari P (2013) Walking pace affected by interactive sounds simulating stepping on different terrains. ACM Trans Appl Percept 10:23

Välimäki V, Pakarinen J, Erkut C, Karjalainen M (2006) Discrete-time modelling of musical instruments. Reports Prog Phys 69(1):1–78

van den Doel K (2004) Physically-based models for liquid sounds. In: Proceedings of the international conference on auditory display (ICAD04), Sydney

van den Doel K (2005) Physically based models for liquid sounds. ACM Trans Appl Percept 2:534–546

van den Doel K, Pai DK (1998) The sounds of physical shapes. Presence Teleoperators Virtual Environ 7(4):382–395

van den Doel K, Pai DK (2004) Modal synthesis for vibrating objects. In: Greenebaum K (ed) Audio anecdotes. AK Peters, Natick van den Doel K, Kry PG, Pai DK (2001) FoleyAutomatic: physically-based sound effects for interactive simulation and animation. In: Proceedings of the international conference on computer graphics and interactive techniques (SIGGRAPH 01), Los Angeles, CA, pp 537–544

Vanderveer NJ (1979) Ecological acoustics: Human perception of environmental sounds. Unpublished doctoral dissertation, Cornell University. (Dissertation Abstracts International, 40, 4543B. (University Microfilms No. 80–04-002))

Velasco C, Jones R, King S, Spence C (2013) The sound of temperature: What information do pouring sounds convey concerning the temperature of a beverage. J Sens Stud 28:335–345

Visell Y, Fontana F, Giordano BL, Nordahl R, Serafin S, Bresin R (2009) Sound design and perception in walking interactions. Int J Hum Comput Stud 67(11):947–959

Voss RF, Clarke J (1975) 1/f noise in speech and music. Nature 258:317–318

Warren WH, Verbrugge RR (1984) Auditory perception of breaking and bouncing events: a case study in ecological acoustics. J Exp Psychol Hum Percept Perform 10:704–712

Waterman NA, Ashby MF (1997) The materials selector, 2nd edn. Chapman and Hall, London

Wildes R, Richards W (1988) Recovering material properties from sound. In: Richards W (ed) Nat Comput. MIT Press, Cambridge, MA, pp 356–363

Yao H-Y, Hayward V, Ellis RE (2005) A tactile enhancement instrument for minimally invasive surgery. Comput Aided Surg 10(4):233–239

Zheng C, James DL (2009) Harmonic fluids. ACM Trans Graph (TOG) 28(3) (Article No. 37)

Zheng C, James DL (2011) Toward high-quality modal contact sound. ACM Trans Graph (TOG) 30(4) (Article No. 38)

Chapter 5
Computational Aspects of Softness Perception

Massimiliano Di Luca and Marc O. Ernst

5.1 Sensory Information About Softness

How do we choose a ripe avocado in a box of seemingly identical ones? How do we test a pillow in a shop to assess whether it will be comfortable once at home? How do we handle a container made of thin plastic, such as an open PET water bottle, that bends in as soon as we put our hand around it? The human brain is able to integrate the unique structural, motor, and sensory properties of our hands and body in order to effortlessly estimate the material properties of deformable objects.

Humans obtain information about material properties through several types of stereotypical manipulations, termed exploratory procedures (Lederman and Klatzky 1996). During these interactions, sensory information about softness is obtained primarily from the tactile and proprioceptive sensory modalities (Tan et al. 1995; Srinivasan and LaMotte 1995). Signals from other sense modalities are also available and can contribute to softness perception. Consider, for example, the vibration produced by aggregate materials when perturbed (i.e., gravel), or the sound produced by the snow, or even the visual change in shape of a pillow when compressed. The vibrotactile, auditory and visual signals can provide information about compliance, at least to some extent. In this chapter, we will analyse the computational principles underlying the combination of multisensory signals during the interaction with deformable objects. Such an analysis facilitates the identification of the information processing mechanism that leads to softness perception.

Deformable objects can have a uniform material or be composed of several parts. In particular, the deformability of the object can differ between its surface and its

M. Di Luca (✉)
Centre for Computational Neuroscience and Cognitive Robotics, School of Psychology,
University of Birmingham, Birmingham, UK
e-mail: m.diluca@bham.ac.uk

M.O. Ernst
Cognitive Neuroscience Department and Cognitive Interaction Technology-Excellence
Cluster, University of Bielefeld, Bielefeld, Germany

© Springer-Verlag London 2014
M. Di Luca (ed.), *Multisensory Softness*, Springer Series on Touch and Haptic Systems,
DOI 10.1007/978-1-4471-6533-0_5

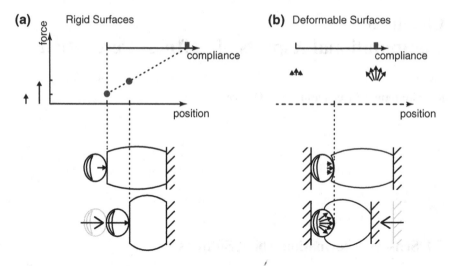

Fig. 5.1 Compliance information with objects having **a** Rigid and **b** Deformable surfaces. **a** Compliance estimates obtained through the combination of position and force information according to Eq. (5.1). The difference in position is divided by the difference in force (see Tan et al. 1995). **b** The compliance of objects with deformable surfaces could be obtained from the pressure map, spread of the contact area and deformation pattern at the contact point, even when position information about indentation position is not available (Srinivasan and LaMotte 1995; Bicchi et al. 2000). That is, tactile information about compliance is obtained in addition to the combination of force and position

interior volume. Springs, for example, are deformable objects that have rigid surfaces (see Chap. 1) so that there is no deformation of local shape at the contact point of the finger, but they are deformable as the external shape changes when force is applied. Such objects are not uncommon: we interact with such spring-like objects with rigid surfaces when we type on a keyboard. In a similar way, force feedback devices can render such spring-like objects, but such devices do not render tactile deformations that are typical of most everyday objects that we interact with. On the other hand, the surface of most compliant objects we encounter is deformable rather than rigid (i.e., pillows, beds, chairs, padded tools, driving wheels). This distinction is paramount as objects with deformable surfaces provide an additional source of information about their compliance, namely the local skin deformation of the finger (Srinivasan and LaMotte 1995). There have been several attempts to create a haptic display that can render tactile information (Chap. 11), and some have even combined tactile and force-feedback displays (Scilingo et al. 2010).

Tactile information for objects with rigid surfaces specifies that the object is not compliant. This information is in conflict with force and position information that, when combined, leads to a different estimate of compliance (Fig. 5.1). The information available during the interaction with objects that have deformable surfaces can be made similar to the case of rigid surfaces, by either providing local anesthesia to the skin area in contact (Srinivasan and LaMotte 1995), or by mediating the contact through a tool (e.g. a pair of pliers). Note that if the perceptual system has access only to force and position information, the brain has no choice but to estimate compliance by combining the force and position signals in the way shown by the formulas below.

Object softness can be quantified for the two types of objects by identifying the amount of stress and the amount of object strain (see Chap. 1). Young's modulus (the ratio between force per unit area and strain) is a measure of stiffness (the inverse of compliance) of objects with soft surfaces. Young's modulus is a property of the material, whereas the stiffness changes with the width of the object. In the case of objects with rigid surfaces, the Young's modulus can be simplified to Hooke's law. Hooke's law states that the change in force (Δf) divided by the displacement Δp is the constant k defined as the object stiffness. The inverse of stiffness k is called compliance C, corresponding toposition difference divided by force change:

$$C = 1/k = \Delta p / \Delta f. \tag{5.1}$$

Compliance is the preferred term in this chapter as it is related to softness.

Several materials obey Hooke's law—i.e., compliance is constant throughout the interaction–as long as the force conditions do not exceed the material's elastic limits (Fig. 5.2a). With non-Hookean materials, instead, force can change as a function of velocity and position, so it is necessary to update the relation between applied force and the change in indentation position in small increments during the exploration of the object (Fig. 5.2b–d). As sensory information for softness perception is available at every instant during interaction, the brain needs to update the perceptual estimate of compliance from the sensory signals available at each time point. As we will discuss below, the time course of sensory signals is important in the perception of material properties as it carries information beyond compliance, e.g., about the rigidity of the object surfaces, about whether the object material is Hookean, or about whether the object is uniform.

5.2 A Bayesian Model of Softness Perception

Different time-varying sensory signals specify the same value of compliance only within some error margin due to noise in the sensory system. That is, the multiple estimates do not necessarily agreeand because of these precision limitations, the observer is faced with the task of inferring environmental properties from imperfect signals. The observer thereby makes a best guess to act upon and will then correct and update the estimate, as well as the action, as soon as more information becomes available. A solution to improve such a process is to combine information obtained from multiple sensory signals and from previous experience in order to predict the state of the world. It is particularly useful to use and combine signals from multiple sense modalities. For the manipulation of objects, useful sources of information can be derived not only from haptics (i.e., proprioception and active touch), but also from other sensory modalities such as passive touch, vision, and audition.

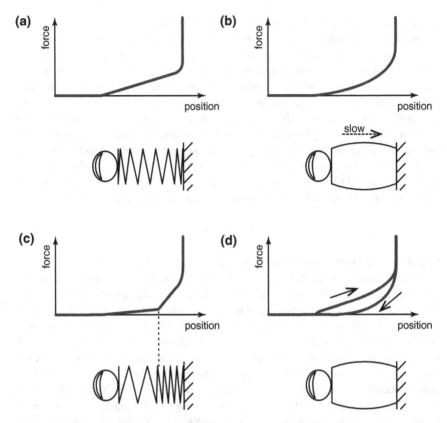

Fig. 5.2 Interaction with different types of objects that can lead to non-Hookean force-position relationships that become evident as deviations from linearity. **a** Complete indentation of an object. **b** Interaction with rubber or silicon specimen, where indentation velocity is low. **c** Composite object (Di Luca 2011) where the more compliant part is indented completely before the stiffer part. **d** Material comprising a damping component where the force during loading and unloading movement does not match

There is an increasing amount of empirical evidence supporting the idea that the brain processes sensory information in a way consistent with Bayesian Decision Theory (Knill and Richards 1996). A Bayesian 'ideal observer' is a mathematical formulation of how perceptual estimates about a property of the world should be obtained and how these estimates should be used to optimally perform a decision (or action) given some cost function.

Bayesian Decision Theory offers a normative way to describe how observers should use information in order to form the most precise and accurate representation of the world, as all processes should be *statistically optimal*. In other words, the goal of the perceptual system should be to reduce uncertainty and, as such, this theory offers a benchmark against which human performance can be compared (Knill and Richards 1996; Mamassian et al. 2002; Ernst 2006). The key for this is the application of Bayes' theorem that uses probability distributions of a physical property in the environment, in this case object compliance.

To use statistically optimal computations like the Bayes' theorem, the brain should combine sensory signals that when taken alone do not carry information about the property of the environment in analysis. Furthermore, the brain should use all available sources of information to obtain a unique estimate of the physical property. Note that following Ernst and Bülthoff (2004), here we define the following terms:

- *Combination* indicates the processing of *complementary* sensory signals, where every component is necessary for an estimate of compliance. Combination of information leads to the likelihood component of the Bayes' formula (see Eq. 5.2). Position and force information, for example, are both necessary in obtaining an estimate of compliance with an object having rigid surfaces (Fig. 5.1a). Combination of information does not lead to an increase in precision of the sensory estimate, as the noise of each of the two variables contributes to the noise of the final estimate.
- *Integration* indicates the use of two or more sensory signals that carry information about the same physical property. In Bayesian terms, each signal can specify a likelihood component of the Bayes' formula (see Eq. 5.2). When interacting with objects that have deformable surfaces, information estimates of compliance can be obtained independently from haptic (using force and position, Fig. 5.1a) and tactile information (Fig. 5.1b). Assuming a relation between the global and the local deformation of the object, the two estimates are *redundant* and the brain can integrate them into one estimate (Fig. 5.3c). This generally will lead to an increase in the precision of the sensory estimate.

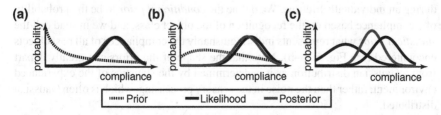

Fig. 5.3 Probabilistic representation of prior, likelihood, and posterior probabilities. **a** Hypothetical *conditional prior* for the class of objects "avocado". Our experience of finding a hard avocado in a box exposed at the vegetable stand is higher than finding a soft one. Note that priors that convey little information (i.e., they tend to be flat) have less influence on perception. **b** Hypothetical *statistical prior* that represents our overall experience with objects in our surroundings representing the fact that is unlikely that we could come in contact with objects with very high or very low compliance, but very low compliance (i.e. hard objects made of metal or wood) is relatively more probable. **c** Maximum Likelihood Estimation with two Gaussian-shaped likelihoods deriving from independent redundant estimates of compliance where prior knowledge is non-influential (i.e. flat, not represented)

With these specifications, the Bayes' theorem states that the probability of a property "a posteriori" is proportional to the likelihood probability multiplied by the "a priori" probability, or:

$$p(property|evidence) \alpha p(evidence|property) \times p(property) \qquad (5.2)$$

When these values are expressed for every value of the environmental property, they can be represented as in Fig. 5.3. Below, we analyze the components involved in this theorem when applied to compliance.

The first component we analyze is the function relating the probability of each possible sensory signal given a value of compliance, which is defined as the *likelihood function*. This is the probability that a state of the world could have generated the sensory signal that is available at the end of an interaction $p(sensory\ evidence|environmental\ property)$. In softness perception, the likelihood function is either obtained from information that directly specifies the compliance of the object (e.g., tactile information for objects with deformable surfaces), or by combining complementary signals (e.g., position and force information for objects with rigid surfaces).

According to the Bayesian theorem above, the likelihood function is combined with prior knowledge about the state of the world $p(environmental\ property)$, which is called the *a priori* probability distribution. The prior represents the statistics of the world with which we interact. For compliance, it represents the probability of encountering a compliance value even before any sensory information is available. In softness perception the prior could either represent the compliance experienced in prior interactions with a category of objects that can be identified visually (i.e. prior to touching it), or it could represent the frequency of compliances encountered during an individual's lifetime. We define the *conditional prior* to be the probability of a compliance based on the recognition of the object class, and we instead call the *statistical prior* what represents indistinguishably the compliances of all past objects encountered (see Fig. 5.3a–b). Note that the shape of the priors can greatly depart from a Gaussian distribution as it is determined by the statistics of the experienced environment, rather than the noise in the sensory processing, which is often Gaussian distributed.

The combination of likelihood function and prior distribution is proportional to the *posterior distribution*, $p(environmental\ property|sensory\ evidence)$. How does perception comes about? From the values of probability expressedat each level of the environmental property to be estimated (Fig. 5.3), the sensation of softness is usually thought to be determined from the maximum of this distribution. Because it is the Maximum of the *A Posteriori* distribution, this is also called the MAP estimate. A desirable result of combining the likelihood with the prior probability is that actions and perception will become more precise and also likely more accurate.

In many cases the prior probability of encountering some property of the world is equally distributed, and thus has little or no influence on the perceptual estimation process. Often, however, we encounter more than one source of sensory information about a particular world property, providing input to the estimation process. Under the assumption of a common cause, likelihood functions obtained from these different sensory signals should be combined into one final estimate of the probability of the state of the world. The application of the Bayesian framework has been very successful in describing how humans integrate redundant sensory information to obtain a more precise percept of the environment. When *a priori* knowledge can be assumed not to exert an influence on perception (i.e., when the range of compliance tested is small and each stimulus is presented an equal amount of times), the prior probability is "flat" and thus it is possible to consider the final estimate as being obtained through the integration of only the likelihood functions.

When the percept is assumed to be determined from the maximum of the combined likelihood function, we call it the maximum likelihood estimate (MLE). In many instances it has been shown that human performance to integrate multisensory signals is close to the statistical optimum according to MLE (Ernst and Banks 2002). With the assumption that the noises of such functions are all Gaussian in shapeand independent, the integrated MLE estimate is a weighted average of the individual uni-sensory estimates, with weights proportional to the inverse variance of the unisensory distributions (the inverse of the variance is the precision and in the cue integration literature is also often termed reliability, Backus and Banks 1999).

The result of MLE is that the integrated estimate is more reliable than either of the two components (Fig. 5.3c). Such a scheme has been applied to the perception of compliance with visual and haptic information about the position of the fingers (Kuschel et al. 2010; Di Luca et al. 2011) and for multiple contact points with the object (Di Luca 2011; Plaisier and Ernst 2012). The perceptual consequences of integration are evident based on the magnitude of the perceptual estimatewhen information in the different modalities contains a small conflict, i.e., when visual and haptic information indicates a different amount of indentation (Kuschel et al. 2010) and when fingers are in contact with an object which has different compliances at the two contact points (Di Luca 2011).

An important outcome of the integration of redundant sources of sensory information is that the uncertainty of the perceptual estimate will be reduced. In order to prove that for compliance perception there is indeed such a reduction in uncertainty, researchers have been comparing performance in discriminating material properties with one and two sources of sensory signals available. If performance with two sources of information is higher than the best single-source, the brain must be taking advantage of the redundancy in the sensory estimates (the signature of sensory integration, Ernst 2006). For compliance perception, the outcome of multimodal integration has been shown to be close to the statistical optimum (maximal reduction in sensory noise) in some cases (Di Luca 2011), but not in others (Kuschel et al. 2010; Cellini et al. 2013). Deviation from optimality seems to occur when conflicts are introduced by the experimental manipulation.

One conflict between redundant sensory signals that can lead to an overweight of information, and can also be ascribed to the detection of discrepancy, is with objects composed of different materials (Bergmann Tiest and Kappers 2009). Information from the deformable surface (sensed through touch) and the overall indentation of the object (sensed by combining force and position) can be made unrelated. With such objects, there is a discrepancy in the compliance estimate sensed from two sensory channels.

A second type of conflict that can lead to non-integration, which does not follow the MLE scheme, is when the visual information about the amount of indentation is manipulated (Kuschel et al. 2010; Cellini et al. 2013). This situation is discussed in detail in Chap. 2. Results suggest that integration uses a fixed set of weights and optimality is limited to the "natural" interaction. Such an outcome could be explained by hypothesising that the conflict is detected, thus increasing the chance that signals are not actually coming from the same external event.

In both cases of conflict (tactile-haptic and visual-haptic), information processing would have to balance the costs of loosing access to the individual estimates against the benefits of improving perceptual precision. Such balance depends on whether the conflict is present at the level of the final sensory estimate of compliance (tactile-haptic) or at the level of the individual complementary force and position components (visual-haptic). In the following section we will evaluate what these two possibilities entail.

5.3 Redundancy in Softness Cues

Integrating sensory information is only beneficial if the estimates are related to the same environmental aspect. Wrongly integrating sensory information leads to inherently erroneous estimates due to conflation. For this reason, integration should occur automatically only once the perceptual system has established the correspondence between estimates (Roskies 1999; Ernst and Bülthoff 2004). It has been shown that temporal and spatial coincidence (Thurlow and Jack 1973), as well as structural similarity between the signals (Parise et al. 2011), is used as an indicator for the perceptual system to consider multimodal information as being related to the same source. Spatial and temporal offset between multisensory signals can, in fact, prevent integration (Witkin et al. 1952; Bresciani et al. 2005). In more complex situations, however, such as during manipulation of deformable objects, spatial and temporal coincidence can be overridden by an inference-like process (Duda et al. 2000) that determines whether sensory signals carry information about the same physical property (Helbig and Ernst 2007).

In the perception of softness there are several ways in which sensory information could be redundant. With the notable exception of touch signal about objects with deformable surfaces (Srinivasan and LaMotte 1995; Bicchi et al. 2000; Fig. 5.1), most sensory signals carry information only related to the indentation or to the force—they do not specify compliance directly. For example, information about the amount of finger movement is carried redundantly by visual and kinesthetic information, but auditory and tactile signals (vibrations and spread of the contact area) also provide some information (contributions of sound is discussed in Chap. 4, vibration in Chap. 3, area spread in Chap. 11). Information about resistive force is carried by touch and proprioception modalities and again other modalities could provide information about force (i.e., vision see Chap. 2 and Cellini et al. 2013). Thus, integration could occur at the level of position and force information, even before a compliance estimate is obtained. The brain could obtain a perceptual estimate of compliance in two ways (see Kuschel et al. 2010):

- Integration of force and/or position estimates (integration before combination). The brain proceeds by first integrating all the position signals into a unique position estimate, and all force signals into one force estimate. Only in a second step are position and force combined to obtain a compliance estimate (Figs. 2.2a, 5.4a, c).
- Integration of compliance estimates (combination before integration). The brain obtains separate estimates of compliance for pairs of signals providing force and position information. In a second step, the various compliance estimates are integrated into a coherent one (Figs. 2.2b, 5.4b, d).

It is still not entirely clear which strategy the brain adopts, and if so, whether it adopts the same strategy under all circumstances. It is possible that the selection depends on which integration strategy is more advantageous in terms of the reliability of the final estimate (see equations below). In an attempt to answer this question we will focus on two examples of interaction with rigid surfaces: First, we will consider the case of proprioceptive and tactile information, where force information is available in each modality (Fig. 5.4a, b), and then we will consider the case of visual and haptic information, where position information is available in both modalities (Fig. 5.4c, d—see Chap. 2 for an in-depth analysis of the visual-haptic case). In the analyses we will assume that the non-indented position of the object and an unloaded null force are used as references for compliance estimates. In such cases, position p and force f could be used in the formulas instead of Δp and Δf. This might not be the case as additional information is available, for example when the distance between the object and the hand is manipulated or when object is made to vibrate orthogonally to the contact point (i.e., Visell et al. 2011). If no other information is available about the location of the first contact, because of the vibration the boundary of the object should be sensed outside the object (i.e., earlier when approaching the object). The sensed indentation difference between null force and peak force should thus appear to be larger. Following this logic, vibration should have the effect of making object appear softer if the boundary is sensed before the interaction takes place (see Chap. 3 for a discussion about vibrotactile information and Chap. 9 for more information about boundary crossing).

5.3.1 Touch and Proprioception

When force information about the indentation of an object with rigid surfaces is conveyed by both proprioceptive and tactile modalities (f_p and f_t), and vision is precluded, the reliability of force (and consequently of compliance), sensed through proprioception, depends on the mechanical configuration and properties of the arm

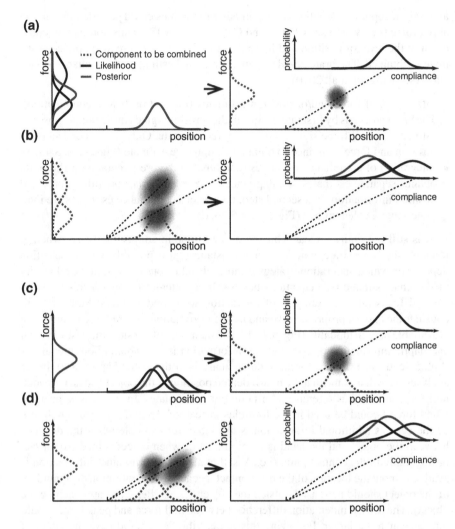

Fig. 5.4 Probabilistic representation of how integration and combination processes could be ordered to obtain a coherent estimate of compliance of objects with rigid surfaces from multiple sensory signals. **a** Integration of force information provided by tactile and proprioceptive modalities (integration before combination). The integration is followed by combination of the posterior estimate of force to obtain a compliance estimate. **b** Combination of force signals with proprioceptive position into separate estimates of compliance (combination before integration). In a second step, the two estimates of compliance are integrated into a coherent one. **c** Integration of position information provided by vision and proprioceptive modalities (integration before combination; here tactile information about force is not represented). The next computation step is the combination of the force and position to obtain a compliance estimate. **d** Combination of force with position information, provided by proprioception and vision, to obtain two compliance estimates which are subsequently integrated (combination before integration)

and hand. Chapter 9 discusses how the reliance on position or force information depends on the distance of the contact point between the participant and the object. Similarly, the results of Di Luca et al. (2011) suggest that the reliability of proprioceptive information can be lowered by making participants use only their shoulder joint to press down on an object rather than using the wrist and elbow. For such reasons, here we will assume that the reliability of proprioceptive information about force r_{fp} can be either higher or lower than the reliability of tactile information r_{ft}. Figure 5.4a shows the condition where the final estimate of compliance is obtained using displacement information and an integrated force difference estimate. In such a case where integration of force according to MLE precedes combination of position and force signal (integration before combination), the final estimate of compliance can be expressed as such:

$$C_{i\_c} = p/f = p/(wf_p + (1 - w)f_t),$$ (5.3)

where the weight is calculated according to reliability of the force estimate r_{fp} and r_{ft} as such

$$w = r_{fp}/(r_{fp} + r_{ft}).$$ (5.4)

The alternative (combination before integration) illustrated in Fig. 5.4b shows two estimates of compliance obtained using a unique proprioceptive position signal. The two compliance estimates are redundant and they can be integrated according to their reliability r_{cp} and r_{ct}. The result can thus be expressed as:

$$C_{c\_i} = wC_p + (1 - w)C_t = w(p/f_p) + (1 - w)(p/f_t).$$ (5.5)

Here the weight w should be calculated according to reliability of the compliance estimates, not the reliability of the force estimates as in the previous case. The difference with this procedure is that both estimates of compliance use the same position information. In this case, the noise affecting the two compliance estimates is not independent. Thus, instead of using the simple formula

$$w = r_{cp}/(r_{cp} + r_{ct}),$$ (5.6)

the weighting scheme should be modified to account for the correlation between noises. The weight needs to be calculated according to the following formula (see Oruç et al. 2003):

$$w = r'_{cp}/(r'_{cp} + r'_{ct}).$$ (5.7)

Here r' is the corrected reliability that accounts for the correlation ρ between the noise in the two compliance estimates and it corresponds to

$$r'_{cp} = r_{cp} - \rho\, r_{cp} r_{ct}$$ (5.8)

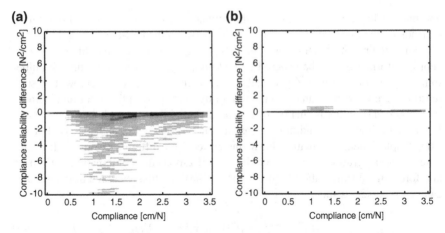

Fig. 5.5 Numerical simulation comparing the reliability of the compliance estimate obtained in 10,000 samples with the two methods in the two scenarios described in Fig. 5.4. Weber Fraction values between 0.1 and 0.5 have been employed for each of the signals involved. Darker color indicates more frequent results. Positive values of the ordinate indicate a more reliable result with a scheme that first combines the signals and then integrates the redundant compliance estimates. If the two methods were mathematically equivalent, we would expect a flat line at 0. **a** Force can be obtained through tactile and proprioceptive modalities, while position is estimated through proprioception. The shaded area below the zero line indicates cases where integration before combination (Eq. 5.3) can produce higher overall reliabilitythan what is obtained with combination before integration (Eq. 5.5). **b** Force is obtained through proprioception and position is obtained redundantly through vision and proprioception. There are very few cases where the reliability is different for the two types of estimates, where the combination before integration (Eq. 5.10) leads to a better result than Eq. (5.12)

and

$$r'_{ct} = r_{ct} - \rho\, r_{cp} r_{ct}. \tag{5.9}$$

We performed numerical simulations to evaluate the reliability of the compliance estimate obtained following Eqs. (5.3) and (5.5). The frequency of the obtained difference in reliability for the two methods is summarized in Fig. 5.5a. Results indicate that there is a demarcated difference between the two methods. At intermediate compliance levels, estimating compliance by firstly integrating redundant force information and only then combining position and force could have higher reliability than the one obtained by firstly combining and then integrating. This is most likely because force estimates obtained with the two modalities have been simulated to differ only moderately. In such cases, the correlation of the noise in the compliance estimates makes the weighting deviate from optimality (Fig. 5.4b) decreasing the reliability of the posterior (Oruç et al. 2003).

5.3.2 Proprioception and Vision

Let us now analyze the case of a compliance estimate obtained with redundant position information provided by vision and proprioception (p_v. and p_p, respectively). Again, this is a case where tactile information does not contribute to compliance estimate directly because contact points are rigid. In such a condition, the visual sense is much more precise than proprioception for estimating object indentation. Redundant position information is integrated before combining it with force to obtain a compliance estimate as shown in Fig. 5.4c. The final estimate of compliance can be expressed by:

$$C_{i\_c} = (wp_v + (1 - w)p_p))/f, \qquad (5.10)$$

where the weight w is calculated according to reliability of the two position estimates r_{pv} and r_{pp} and thus it is expected to be high for vision:

$$w = r_{pv}/(r_{pv} + r_{pp}). \qquad (5.11)$$

On the other hand, integrating redundant estimates of compliance after position and force have been combined as shown in Fig. 5.4c would lead to a compliant estimate equal to

$$C_{c\_i} = w\,C_v + (1 - w)C_p = w\,(p_v/f) + (1 - w)(p_p/f), \qquad (5.12)$$

but again the weight is not calculated according to the simple reliabilities

$$w = r_{cv}/(r_{cv} + r_{cp}), \qquad (5.13)$$

It is computed according to the corrected reliability values shown above, which consider that noise is not independent

$$w = r'_{cv}/(r'_{cv} + r'_{cp}), \qquad (5.14)$$

thereby taking into account the correlation between the noise of the estimates due to the use of the same force signal. The results of the numerical simulation displayed in Fig. 5.5b show the difference in reliability between compliance obtained following Eqs. (5.10) and (5.12). By accounting for the correlation between variances of the estimates, there are small differences in the reliability attainable with the two methods, with a modest improvement in performance by first combining information in separate compliance estimate, and then integrating them. This is the result of improvement in precision due to the presence of two estimates of compliance. Such improvement is lost when integrating position information because of the extreme weight that should be assigned to the visual estimate of position.

Results of the simulation suggest that integrating force information to obtain a unique force estimate to be combined with position information might lead to a more precise overall compliance estimate. On the other hand, integrating position information available through proprioception and vision leads instead to a slightly worse final estimate of compliance than first obtaining two separate estimates of compliance to be integrated.

The finding that the brain is better off by combining force and position signals when visual and proprioceptive modalities are available is consistent with the finding of Kuschel et al. (2010). The researchers found that compliance reliability is very low when only visual position information and haptic force information are available. Such combination of sensory information is not sufficient to estimate compliance if the brain first computes a compliance estimate and then integrates the various estimates available (see Chap. 2 for an in depth discussion). It would be interesting to perform a similar test with touch and proprioceptive information (force information available through touch, but not through proprioception), as for this case the best performance can be achieved by first integrating force signals and only then combine with position (Fig. 5.5a).

To summarize, integrating compliance information, rather than position or force information, should depend on the correlation between the noise sources affecting the sensory signals involved in the process. One could ask whether such a difference is also present when multiple spatially-separated sensors are available (i.e., when interaction with an object is performed with multiple fingers). Such a case will be considered in the next section.

5.4 Multiple Contact Points

Our perception of the compliance seems to depend on the way we interact with of objects (Di Luca 2011; Kaim and Drewing 2009). The overall judgment of the quality of the fruit depends on the active exploration and the pattern of movement of the fingers. So how much movement do we plan to apply to begin with? Chap. 6 shows how, for example, the amount of force depends on the difficulty of the anticipated discrimination, and on the compliance of the object that is expected, as using the appropriate force would increase the discrimination sensitivity. Moreover, with very compliant objects better performance can be achieved using 3 fingers rather than only 1 (see Chap. 6). Another question that has not been addressed is whether participants *choose* whether to use one or multiple fingers in the interaction and when do they decide to switch. Finally, it is not clear whether using multiple fingers in contact with compliant objects leads to a specialized role of each finger. Some contacts might be used predominantly for stabilisation while other might be used for active exploration. In extreme cases when one presses down on an object with just one finger there would be only one contact point that provides information about compliance. Another way to assess the softness of the object would be to grasp it with two fingers in a "pinch" grip. Employing such a precision grasp could gather information about material properties from both contact points. Each contact point is one sensor that collects sensory information that needs to be integrated appropriately. But the information they collect could vary for example if the grip is horizontal or vertical.

There are also cases when compliance information is obtained with a full hand grasp, or even two hands. Here, the information from each finger and contact point can be used as a separate estimate of the overall compliance of the object. In such cases of multiple contacts, we need to consider that compliance is not constant across the object, i.e., fruit might have some areas where the material is soft because it starts to rot. Thus, with multiple contact points, the spatially separated estimates need to be integrated into one percept of the object compliance, assuming an object with compliance properties that are similar throughout its surface.

Many studies investigating compliance perception with objects having rigid surfaces mostly analyzed single contact-point interactions (Srinivasan and LaMotte 1995; Di Luca et al. 2011). The few cases that analyzed more than one contact point, as it happens in the case of object holding (Chen and Srinivasan 1998) or grasping (Roland and Ladegaard-Pedersen 1977; Tan et al. 1995; Freyberger and Färber 2006; Kuschel et al. 2010; Di Luca 2011), considered the sensory information as the sum of the two forces and deformations. If the object is squeezed between fingers, i.e., contact points are producing forces in opposite directions, one could assume that forces and deformations at the two ends should be simply summed up. Such assumption, however, is not always fulfilled, for example, as in the case where objects are pressed down with two fingers. Moreover, by considering that with pinch grasps it is possible to generate independent finger movements (Schieber 1996; Smeets and Brenner 2001), it becomes apparent that the fingers can act as separate sensors to collect information about resistive force and deformation. A simple sum of the two sources of information would not be statistically optimal in all situations (it would not lead to the most precise estimate of compliance). To obtain an optimal estimate of compliance from two sensory sources, each source should be weighted according to its reliability, consistent with Bayesian inference (Knill and Richards 1996). Indeed it has been shown that in many instances the perceptual system is, in principle, capable of obtaining close to optimal performance in compliance judgments obtained with multiple contact points (Di Luca 2011; Plaisier and Ernst 2012). The weighting of each contact estimate changes as a function of the exploratory movement reflecting the reliability of the estimate (Di Luca 2011). This weighting scheme is consistent with the relative reliability of compliance estimates when the object is composed of a uniform material. Performance obtained in this case is close to optimal. Others have found a close to optimal performance in the case of consistent sources of compliance information (Kuschel et al. 2010; Di Luca et al. 2010).

The perceptual system should only integrate information that comes from the same distal source and should behave in a robust way otherwise. That is, when conflicts or evidence about the origin of sensory information indicate the presence of separate sources, integration should simply not occur. For example, weighted averaging should not occur mandatorily when large conflicts are present between different sources of information (i.e., van Ee et al. 2002). Di Luca (2011) also finds that perceived compliance is mostly dictated by one source of information when the conflict is large, however the source chosen to drive the percept is not the one that provides more reliable information. Different results show, in fact, that there is a lawful relation between reliability and overall compliance (Jones and Hunter 1990;

Tan et al. 1995). If only compliance judgment reliability was at stake, there should be a consistent weighting according to the reliability difference. Instead, results by Di Luca (2011) suggest that when using a precision grip, it is the information coming from the finger that moves the most that drives the percept. Work in other domains also showed that the source of sensory information chosen to drive the percept was not necessarily the more reliable one (Girshick and Banks 2009; Gori et al. 2008). Overall these results are in line with a scheme of integration that can lead to near-optimal perception, but beyond the limit of fusion progressively increases the weighting of one source of information. It does not, however do this based on reliability alone.

5.5 Temporal Aspects of Softness Perception

Whether the computation proceeds by first integrating and then combining, or by first combining and then integrating; either way, force and position information are necessary to obtain a softness percept of objects with rigid surfaces (Fig. 5.1a). Several studies investigating compliance perception of objects with rigid surfaces make two major assumptions:

- First, that the objects to be manipulated are in "unloaded" resting state, so that even infinitesimal forces can create an indentation albeit infinitesimal. See Tan et al. (1995) for an exception.
- The second premise is that one of the two indentation positions considered for the estimate of compliance is the resting state of the object (the non-indented position) and thus, only one position is analysed (often the one at the maximum indentation).

These simplifications offer the advantage of reducing the problem of estimating compliance to the estimate of a single force and a single position, seemingly obtained at the point where force reaches the maximum (see Tan et al. 1995 for an investigation). These premises, however, are not always fulfilled and create computational problems when dealing with non-Hookean materials. For example, if force and position are related to the starting point of indentation (i.e., the point at which the first contact is made with the objects surface), then the non-indented position of the object should have a paramount influence on perception (see Nisky and Mussa-Ivaldi 2008). Pressman et al. (2011) indeed showed that the unnoticed movement of the object could cause a misperception of compliance that can be ascribed to the observer using previous estimates of the non-indented position. This information would act as a prior for position to be integrated at the force/position level. Interestingly, small force gradients of very compliant objects could lead to biases in the estimation of the perceived location of the non-indented position. Namely, the perceived location of the non-indented surface is biased inwards due to sensory thresholds for force (Chap. 7 includes a detailed analysis of the influence of the perceived location of the boundary of a force field). In other words, if compliance is estimated only through the haptic modality, and information using the non-indented position of the object is used as a reference, perceived softness with very compliant objects should be lower than veridical (i.e., very compliant objects should be perceived less compliant than they are). Chapter 9 analyses the case of the interaction with haptic interfaces where force rendering is delayed with respect to position. With a delay in the force generated by the device, the perceived position of the non-indented object should be biased to a position further inside the object (Fig. 5.3d). Consequently, the slope used to estimate compliance (Fig. 5.1a) should be higher, leading to a less compliant percept. Interestingly, there is an indication that with haptic-only interaction, a delay in the rendering of force leads to underestimation of compliance–objects are perceived to be harder when a force delay is present (Pressman et al. 2007). On the contrary, adding visual information about the indentation, in particular about the position of the non-indented surface of the object, reverses the effect making force-delayed objects appear more compliant than non-delayed ones (Ohnishi and Mochizuki 2007; Di Luca et al. 2011).

It is important to note that integration of compliance information allows the use of multiple estimates, which has the advantage of improving reliability. A positive correlation between the noises corrupting the sensory information, however, leads to a reduction of the overall reliability of the final, integrated estimate of compliance (i.e. Oruç et al. 2003). Moreover, integrating redundant compliance information, rather than redundant position or force, allows the brain to use prior knowledge about compliance in the final estimate. Prior knowledge about compliance is independent of the actual force and position values. It is known that knowledge about the regularities of the world can help to reduce uncertainty and ambiguity in perception by creating an *a priori* expectation of what the most probable state of the world is (Knill and Richards 1996). The knowledge acquired from past experience interacting with the object, e.g., about its compliance, can be expressed using a probability distribution across all possible compliances. *A priori* information has a strong effect on perception in everyday situations. Namely, priors actually influence many of the properties of the perceptual world and have a substantial impact on the dynamics of sensory processing. In particular, priors act as a predictor for the state of the world for which motor actions are initially planned. When approaching an object in order to sense its softness, the brain needs to have an approximate idea of what the softness of said object is, so that actions after contact are executed accordingly. Actions are planned on the basis of *a priori* expectations about the material properties. Such internal models based on previous interactions (Kawato et al. 1987) are triggered from the view of the object before contact. In other words, recognizing an object will activate the internal model and create an expectation of what it will feel like when in contact. This expectation should be used for both perception and action. We discussed the influence of priors on perception, but what happens on action could be resumed by considering the example situation where we expect the object to be very soft. Here we should NOT approach the object by applying a large force (i.e., a strong grip force), as this will compress the object entirely. We should instead approach the object using a low force to perturb it, so as to increase the precision of force and position signals used to estimate compliance. In other words, the best strategy to improve the precision of a perceptual judgment is to base the movement on prior knowledge about how the object will react, so to detect any discrepancy in compliance from the initial guess.

Another problem to overcome is that there is a delay between the motor command and the physical changes in the environment that will generate sensory information. This constitutes a computational problem that could be solved using a forward model about the motor action (Miall and Wolpert 1996). This model should also be updated if it is not accurate when new sensory information comes about. In other words, as sensory signals become available after contact, the influence of the internal model of the material (prior knowledge) should decrease and discrepancies between the forward models and the actual state of the world should be used to update the estimate, as well as to change the movement parameters.

What complicates things even further is that the estimate of compliance of objects having rigid surfaces requires an active indentation of the object (i.e., a *change* in the global deformation of the object through the application of a different force). Because of this, the estimate of compliance undergoes continuous updating as more sensory information becomes available over time. The Kalman filter is a statistically optimal method that can update estimates over time. Human performance has been hypothesized to employ computational mechanisms similar to the Kalman filter in sensorimotor integration (Wolpert 1997) and in perception (Rao 1999), but it is not clear whether such a statistically optimal update happens to the internal model of compliance.

Notably, as the estimate of compliance is dynamic, it is not possible to know experimentally what the perception of softness is at any *one* moment during the interaction. Experiments have usually employed the task of judging the object compliance after the interaction has been completed. Similarly, the concept of minimizing uncertainty over the course of the exploration cannot be extended directly to every time point during the interaction. The series of movements needed to first reach, grasp and then probe a deformable object involves specific cost functions, and each has an accumulation of sensory information with a different goal. For example, parts of the task, like reaching and grasping, might involve minimizing energy consumption or execution time, while other components of the task could consider the chances of hitting other objects. In such cases, looking at overall performance and characterising it as noise reduction in the response to an experimental task might not be the best way. Information accumulation could be better characterized with an analysis of information available and movement performed at different time points over the course of the interaction (Fig. 5.6). For example, when estimating the size of a bar using touch and vision we accumulate tactile information over the first second, while the time course is much faster for vision (see Ernst 2001). Thus, longer trial durations would result in a more precise size estimate through the haptic information, but no comparable increase should be present for visual information. In the same way, during interactions with compliant objects the quality of haptic sensory information can increase with more prolonged interactions. The situation is further complicated as compliance perception requires movement (i.e., information is gathered through motion, whether this is planned and self-generated, or not).

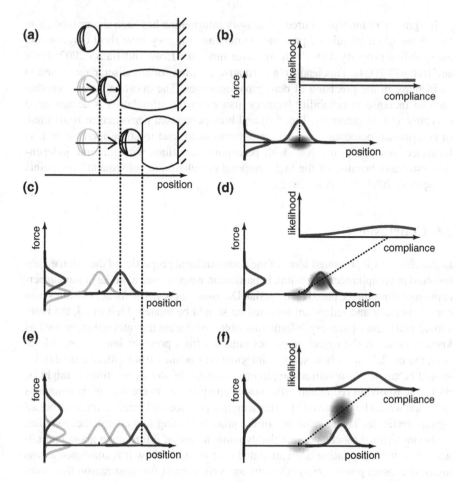

Fig. 5.6 Representation of the interaction with a soft object over time. **a** Force and position sampling distributions. **b** The reliability of the compliance estimate increases as more sampling points become available

Di Luca et al. (2011) have shown that the phases of the indenting action (loading the object with a force and unloading it) are differently informative regarding compliance. Namely, unloading actions are not as informative about compliance as loading ones. You could demonstrate this for yourself by comparing the availability and strength of a softness percept while pushing against an object and releasing the applied force. This means that when an object is pushed against, there can be many sensory signals on which the movement planning can be optimized upon (amount of deformation over time with correspondingly resistive force values, vibratory and auditory patterns, skin pressure distribution, stretching dynamics, etc.). Movements can thus change after each repetition so to increase the reliability of the final estimated property.

Integration of multiple sources of sensory information has been shown to be close to optimal when stimuli are presented simultaneously,they have short duration and the specified property doesn't change over time (see Ernst and Banks 2002; Ernst and Bülthoff 2004). This leads to a normative solution for any discrepancies and to an increase in the precision of perceptual estimates. The question here is whether such an increase in reliability happens also when the stimulation is accumulated over time (i.e. sequentially, Fig. 5.6) and how perceptual performance is affected. In compliance perception we expect integration to lead to some increase in performance, but not as much as with perceptual situations with multiple independent estimates because of the high temporal correlation at subsequent time points (Oruç et al. 2003; Juni et al. 2012).

5.6 Conclusions

In this chapter we presented some of the computational properties of the mechanisms involved in compliance perception. In particular, we proposed a model of human perception of compliance based on Bayesian Decision Theory. The model indicates how complementary and redundant information should be treated. First of all, the brain should make use of sensory information obtained during the interaction, as well as knowledge about the object properties obtained from previous interactions. Moreover, the model shows how sensory information obtained from different modalities should be treated to obtain a compliance estimate. In some conditions it might be profitable to obtain redundant estimates of compliance, whereas in other situations precision would be improved by using a unique position and force estimates from all signals available. The adaptability of the process leading to a compliance estimate is also underlined by considering the dynamic nature of information about compliance, how the information is accumulated over time, and how it is combined across multiple contact points. Results of this analysis suggest that one reason that compliance is normally perceived through multiple sensory signals, across several sense modalities, and over time is that such a range of signals leads to an improvement of the quality of the final compliance estimate. Such improvement is most important if direct contact with the object does not provide direct information about its softness, as is the case when interacting with objects having rigid surfaces.

Acknowledgments The authors are grateful to Markus Rank and Darren Rhodes for help in preparing the manuscript.

References

Backus BT, Banks MS (1999) Estimator reliability and distance scaling in stereoscopic slant perception. Perception 28:217–242
Bergmann Tiest WM, Kappers AML (2009) Cues for haptic perception of compliance. IEEE Trans Haptics 2(4):189–199

Bicchi A, Scilingo EP, De Rossi D (2000) Haptic discrimination of softness in teleoperation: the role of the contact area spread rate. IEEE Trans Robot Autom 16(5):496–504

Bresciani J-P, Ernst MO, Drewing K, Bouyer G, Maury V, Kheddar A (2005) Feeling what you hear: auditory signals can modulate tactile taps perception. Exp Brain Res 162(2):172–180

Cellini C, Kaim L, Drewing K (2013) Visual and haptic integration in the estimation of softness of deformable objects. I-Perception 4(8):516–531

Chen JS, Srinivasan MA (1998) Human haptic interaction with soft objects: discriminability, force control, and contact visualization. MIT, RLE technical report 619 (pp 1–208)

Di Luca M (2011) Perceived compliance in a pinch. Vis Res 51(8):961–967

Di Luca M, Knorlein B, Ernst MO, Harders M (2011) Effects of visual-haptic asynchronies and loading-unloading movements on compliance perception. Brain Res Bull 85(5):245–259

Duda RO, Hart PE, Stork DG (2000) Pattern classification, 2nd edn. Wiley, New York

Ernst MO (2001) Psychophysikalische Untersuchungen zur visuomotorischen Integration beim Menschen: visuelle und haptische Wahrnehmung virtueller und realer Objekte. Eberhard-Karls-Universität Tübingen

Ernst MO (2006) A Bayesian view on multimodal cue integration. In: Knoblich G, Thornton IM, Grosjean M, Shiffrar M (eds) Human body perception from the inside out. Oxford University Press, NY, pp 105–131

Ernst MO, Banks MS (2002) Humans integrate visual and haptic information in a statistically optimal fashion. Nature 415(6870):429–33

Ernst MO, Bülthoff HH (2004) Merging the senses into a robust percept. Trends Cogn Sci 8(4): 162–169

Freyberger FKB, Färber B (2006) Compliance discrimination of deformable objects by squeezing with one and two fingers. Proc EuroHaptics 06:271–276

Girshick AR, Banks MS (2009) Probabilistic combination of slant information: weighted averaging and robustness as optimal percepts. J Vis 9(9):1–20

Gori M, Del Viva M, Sandini G, Burr DC (2008) Young children do not integrate visual and haptic form information. Curr Biol 18(9):694–698

Helbig HB, Ernst MO (2007) Knowledge about a common source can promote visual haptic integration. Perception 36:1523–1533

Jones LA, Hunter IW (1990) A perceptual analysis of stiffness. Exp Brain Res 79:150–156

Juni MZ, Gureckis TM, Maloney LT (2012) Effective integration of serially presented stochastic cues. J Vis 12(8):12–12

Kawato M, Furukawa K, Suzuki R (1987) A hierarchical neural-network model for control and learning of voluntary movement. Biol Cybern 57(3):169–185

Kaim L, Drewing K (2009) Finger force of exploratory movements is adapted to the compliance of deformable objects. In: Proceedings world haptics 2009, third joint EuroHaptics conference and symposium on haptic interfaces for virtual environment and teleoperator systems. The Institute of Electrical and Electronics Engineers (IEEE), Piscataway, NJ, pp 565–569

Knill DC, Richards W (1996) Perception as Bayesian inference. Cambridge University Press, Cambridge

Kuschel M, Di Luca M, Buss M, Klatzky RL (2010) Combination and integration in the perception of visual-haptic compliance information. IEEE Trans Haptics 3(4):234–244

Mamassian P, Landy MS, Maloney LT (2002) Bayesian modelling of visual perception. In: Rao R, Olshausen BA, Lewicki MS (eds) Probabilistic models of the brain: perception and neural function. MIT Press, Cambridge, pp 13–36

Miall RC, Wolpert DM (1996) Forward models for physiological motor control. Neural Netw 9(8):1265–1279

Nisky I, Mussa-Ivaldi FA (2008) A regression and boundary-crossing-based model for the perception of delayed stiffness. IEEE Trans Haptics 1(2):73–83

Ohnishi H, Mochizuki K (2007) Effect of delay of feedback force on perception of elastic force: a psychophysical approach. IEICE Trans Commun E90-B(1):12–20

Oruç İ, Maloney LT, Landy MS (2003) Weighted linear cue combination with possibly correlated error. Vis Res 43(23):2451–2468

Parise CV, Spence C, Ernst MO (2011) When correlation implies causation in multisensory integration. Curr Biol 1–4

Plaisier MA, Ernst MO (2012) Two hands perceive better than one. In: Haptics: perception, devices, mobility, and communication. Lecture notes in computer science, vol 7283, pp 127–132

Pressman A, Karniel A, Mussa-Ivaldi FA (2011) How soft is that pillow? The perceptual localization of the hand and the haptic assessment of contact rigidity. J Neurosci 31(17):6595–6604

Roland PE, Ladegaard-Pedersen H (1977) A quantitative analysis of sensations of tensions and of kinaesthesia in man. Brain 100:671–692

Rao RP (1999) An optimal estimation approach to visual perception and learning. Vis Res 39(11):1963–1989

Roskies AL (1999) The binding problem. Neuron 24:7–9

Schieber MH (1996) Individuated finger movements: rejecting the labeled-line hypothesis. In: Wing AM, Haggard P, Flanagan JR (eds) Hand and brain. Academic Press, New York, pp 81–98

Scilingo EP, Bianchi M, Grioli G, Bicchi A (2010) Rendering softness: integration of kinesthetic and cutaneous information in a haptic device. IEEE Trans Haptics 3(2):109–118

Smeets JBJ, Brenner E (2001) Independent movements of the digits in grasping. Exp Brain Res 139:92–100

Srinivasan MA, LaMotte RH (1995) Tactual discrimination of softness. J Neurophysiol 73(1): 88–101

Tan HZ, Durlach NI, Beauregard GL, Srinivasan MA (1995) Manual discrimination of compliance using active pinch grasp: the roles of force and work cues. Percept Psychophysics 57(4):495–510

Thurlow WR, Jack CE (1973) Certain determinants of the "ventriloquist effect". Percept Mot Skills 36:1171–1184

van Ee R, van Dam LCJ, Erkelens CJ (2002) Bi-stability in perceived slant when binocular disparity and monocular perspective specify different slants. J Vis 2(9):2–2

Visell Y, Giordano BL, Millet G, Cooperstock JR (2011) Vibration influences haptic perception of surface compliance during walking. PloS One 6(3):e17697

Witkin HA, Wapner S, Leventhal TJ (1952) Sound localization with conflicting visual and auditory cues. J Exp Psychol 43:58–67

Wolpert DM (1997) Computational approaches to motor control. Trends Cogn Sci 1(6):209–216

Part II
Sensorimotor Softness

Chapter 6
Exploratory Movement Strategies in Softness Perception

Knut Drewing

6.1 Introduction

Perception is an active process during which humans purposively gather sensory information in order to obtain a representation of their environment. Haptic perception is a prime example of this principle (Gibson 1962). When humans aim to perceive their environment by touch, for example when they aim to haptically perceive an object's softness, they first need to appropriately explore the object with the fingers. It is this exploratory movement that generates the relevant sensory information. This chapter deals with the exploratory movements that humans execute when they aim to judge an object's compliance. Compliance is a physical correlate of perceived softness that is defined as the inverse of stiffness and can be considered a surface's "resistance" to deformation. In the simplest case (cf. Chap. 1) it can be measured as an object's deformation in response to an applied force, e.g. in (milli) meters per Newton.

6.2 Exploratory Procedures in Softness Perception

6.2.1 Pressure

In active touch, humans systematically use different movement schemes or patterns of contact to perceive different haptic properties: the so-called Exploratory Procedures (EPs, see Klatzky and Lederman 1999; Klatzky et al. 1989; Lederman and Klatzky 1987). For example, to perceive texture humans typically produce

K. Drewing (✉)
Institute for Psychology, Giessen University, Giessen, Germany
e-mail: knut.drewing@psychol.uni-giessen.de

© Springer-Verlag London 2014   109
M. Di Luca (ed.), *Multisensory Softness*, Springer Series on Touch and Haptic Systems,
DOI 10.1007/978-1-4471-6533-0_6

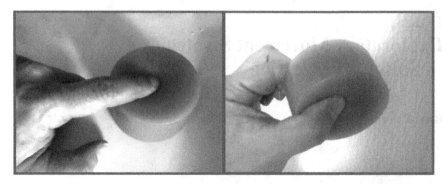

Fig. 6.1 Exploratory Procedure *pressure* as typically executed (*left side*) by pressing with a single finger into an object that is supported by a *table* and (*right side*) by a pinch grip using *thumb* and *index finger*

shear forces between an object and the skin (EP: lateral motion), or to perceive temperature humans maximize the contact area between skin and object without moving (EP: *static contact*). Humans stereotypically and habitually use specific EPs in association with specific properties. The EP used during compliance judgments has been called *pressure*:

> *Pressure*: associated with encoding of compliance; characterized by application of forces to object (usually, normal to surface), while counterforces are exerted (by person or external support) to maintain its position; (Klatzky and Lederman 1999; p. 172)

Humans, typically, execute this EP by pressing the finger pad into the surface or by lifting the object in a pinch grip and squeezing it between thumb and index finger (Fig. 6.1). However, poking, tapping, or twisting movements can also occur (Lederman and Klatzky 1990). Often, the explored objects are pressed a number of times in succession. That is, when the fingers are in contact with the object, normal forces applied to the surface follow a pattern of increases and decreases over time (Kaim and Drewing 2011; Lederman and Klatzky 1987). Figure 6.2 shows part of the typical variation in finger forces while exploring a deformable rubber stimulus for the purpose of compliance discrimination, where only the index finger is used. In the depicted trial, after the first finger-stimulus contact, the force recorded remained at a moderate level for the following 90 ms. After this, the force increased and the finger started to deform the stimulus. In this example, there were two force maxima (ramping forces that peak) at 370 ms after first contact (maximum initial force) and 690 ms after first contact. Individuals will often make multiple indentation movements and in this experiment, another 1–2 further force maxima followed (Kaim and Drewing 2011; Exp. 1).

The stereotypical association of the EP *pressure* with compliance judgments, as well as the stereotypical associations between other EPs and haptic properties, has been demonstrated in several tasks. For instance, when participants have to match or sort objects according to perceived softness, they use the EP *pressure*, whereas

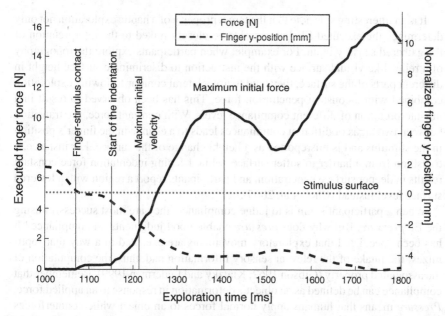

Fig. 6.2 Part of a typical finger force variation while exploring a deformable rubber stimulus. Depicted are the force applied by the index finger to the stimulus (*solid line, left ordinate*) and the corresponding vertical finger position (*dashed line, right ordinate*). Position 0 corresponds to the stimulus surface; negative values of position represent an indentation of the stimulus. In the depicted trial, the finger contacted the stimulus ca. 1,070 ms after trial initiation. Roughly 90 ms after first contact, the force increased with time as the finger indents the stimulus. Depicted is a trial where two indentation movements have been executed with peaks at 1,400 and 1,760 ms (data from Kaim and Drewing 2011)

they will use *static contact* for temperature judgments, *lateral motion* for texture judgments, and so on (Klatzky et al. 1989; Lederman and Klatzky 1987). Similarly, in haptic object identification, after having lifted the object, participants use the EP associated with the property that is diagnostic for the specific task; for instance, for the specific question "Is this bread a stale bread?", compliance is diagnostic and the EP *pressure* is used (Lederman and Klatzky 1990). Additionally, it has been shown that the stereotypical EPs are superior to other EPs in perceiving the associated haptic property: Participants who were constrained to use a specific EP when matching objects according to a specific haptic property performed best with the EP that is stereotypically used for this property (Lederman and Klatzky 1987, Exp. 2; Lederman and Klatzky 1993). In these experiments, compliance matches obtained with the EP *pressure* turned out to be significantly more accurate than when using any other of the six tested EPs (*pressure, static contact, lateral motion, contour following, enclosure, unsupported holding*). Some information on compliance was also available using the other EPs, but the EP *pressure* itself tends to be highly specialized for compliance.

It is an interesting side note that the target property of a haptic exploration not only determines the executed EP, but can also be closely linked to the interpretation of the gathered sensory input. For example, when participants explore the topography of a relief-like virtual surface with the instruction to discriminate surface height in different parts of the surface, they tend to use a lateral contour following exploration combined with a constant penetration force. This has been observed to result in a misinterpretation of different compliance levels. With constant force, the transition between two areas of different compliances leads to a change in the finger's position in the stimulus and is misreported as a height change of the surface. For instance, a transition from a harder to softer surface, while keeping indentation force constant results in deeper surface penetration, and participants report a region with a lowered surface level instead (Choi et al. 2005).

When a participant's aim is to judge compliance, they are most successful using the EP *pressure*. But why does *pressure* enable good judgments on compliance? It has been speculated that exploratory movements are executed in a way that 'optimizes' the intake of the relevant sensory information and, thus, the computation of the associated property (Gibson 1962; Klatzky and Lederman 1999). Remember that compliance can be defined as an object's deformation in response to an applied force. *Pressure* means that humans apply normal forces to an object while counterforces are exerted and the object is deformed. Kinaesthetic signals from muscles, tendons, and perhaps joints provide information about the finger position and finger force, while tactile signals are obtained from the mechanoreceptors in the skin of the finger pad and can provide information about the deformation of the finger-stimulus contact area (e.g., Bicchi et al. 2005). With the EP *pressure*, an object's compliance can be judged from the relation between the applied or sensed forces and the sensed object deformation. That is, the EP *pressure* enables good judgments on compliance, because it generates crucial sensory signals and also because it generates redundant tactile and kinaesthetic signals which humans can integrate in order to assess compliance (Bergmann Tiest and Kappers 2008; Srinivasan and LaMotte 1995).

6.2.2 Tapping as an Instance of Pressure

So far, this section has dealt with the typical execution of the EP *pressure,* which is performed by pressing one or two bare fingers several times relatively slowly into an object. Humans are also able to use tapping movements in order to estimate softness (LaMotte 2000). Tapping movements are another example of the EP *pressure*. Tapping movements are, however, faster, briefer and less forceful than pressing movements and can be performed with the bare finger or with a tool, i.e. a stylus (Fig. 6.3).

LaMotte (2000) investigated how well humans discriminate two objects of different compliances (range: 0.2–2.2 mm/N) with tapping as compared to pressing movements, both in bare-finger and stylus conditions. They found slightly better discrimination performance with tapping than with the slower pressing movements.

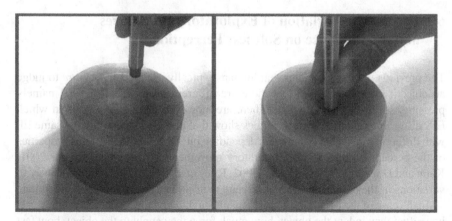

Fig. 6.3 Tapping with a stylus. Here, the stylus or tool is held with two fingers and quickly tapped onto the stimulus. Tapping movements can also be executed with the bare finger. In both cases they allow for highly successful judgments of stimulus compliance

Friedman and colleagues (2008) asked participants to estimate the softness of a broad range of objects (compliances: 0.2–7.5 mm/N) using the method of magnitude estimation (cf. Chap. 1) and their participants' softness estimates were comparable for bare-finger pressing and bare-finger tapping. Overall, judgments on softness are similar, independent of whether humans used tapping or pressing movements to explore the objects.

LaMotte (2000), however, also showed that different sensory signals are crucial for judging compliance when using tapping as compared to the use of pressing movements. He compared compliance discrimination when using tapping and pressing movements in active touch versus passive touch. In active conditions, participants actively moved a stylus with the finger in order to judge compliance. In contrast, in passive conditions the finger was stationary and did not move, while it received similar stimulation from the stylus as in the active conditions. As already mentioned, compliance discrimination was similar for both active tapping and active pressing. In passive conditions, however, tapping movements allowed for compliance discrimination, while pressing movements did not. This finding demonstrates that both active and passive tapping movements generate sensory signals that are not available from pressing movements. In particular, LaMotte (2000) suggested that in the fast tapping movement, the high speed contact with an object provides important information on compliance. Other authors also emphasize a potential role for these "impact tactile cues" in softness perception (Friedman et al. 2008; Lawrence et al. 2000). For a review of such findings and a discussion of the contribution of vibrotactile information see Chap. 3.

6.3 Parametric Variation of Exploratory Procedures and the Influence on Softness Perception

The previous section illustrates that humans typically use the EP *pressure* to judge an object's compliance, and that there are different instances of this EP, namely pressing, squeezing, and tapping. There are, however, more subtle ways in which EPs differ. In particular, several studies show that participants execute the same EP with different movement parameters depending on, for example, stimulus properties or the perceptual aims of the exploratory movements (Gamzu and Ahissar 2001; Kaim and Drewing 2008; Nefs et al. 2002; Riley et al. 2002). This section deals with variations of the parameters of *pressure* and their influences on softness perception.

In the case of the EP *pressure*, participants might vary how many fingers they use, how often they indent the object, how much force they apply to the object, how fast they build-up the force, how they orient their fingers relative to the object, and so on. Also, the distinctions between the instances of the EP *pressure* can be described by their movement variations. Tapping and pressing movements, for example, differ in the rate of force change over time, with force changing faster in tapping than in pressing. Tapping and pressing further differ in the peak forces, which are lower in tapping (Friedman et al. 2008; LaMotte 2000).

Differences in movement parameters may affect how well humans perceive softness (i.e., applying pressure to the object leads to better softness discrimination performance than releasing it, Di Luca et al. 2011). Additionally, the relation between movement parameters and perception may be modified by characteristics of the to-be-explored stimulus, such as its compliance or the type of its surface (which can be deformable like rubber or rigid like a piano key; Srinivasan and LaMotte 1995). This section will specifically focus on movement parameters that are likely to be directly controlled for by the participant, and thus are motor parameters in a proper sense. These include force-related parameters, because the motor system directly controls the force output of the muscles (cf. Kaim and Drewing 2011) and the number of fingers being used. Note that other EP parameters, such as the finger position, are therefore better regarded as a result of controlled motor parameters and potentially provide sensory cues to an object's compliance. For example, in softness perception, if participants have direct control of the forces exerted (which would be the case in active-touch paradigms), then the finger's position in the object will be a function of both force and object compliance, and thus, provide a sensory cue to compliance. It is also possible that particular constrained task situations can effectively modify the roles of different EP parameters as motor parameters or sensory cues.

6.3.1 Peak Forces

Several studies considered how the use of different maximal or *peak forces* during stimulus deformation affects softness perception (cf. Fig. 6.2; Friedman et al.

2008; Tan et al. 1993). Srinivasan and LaMotte (1995) were among the first who investigated the influence of applied peak force on compliance discrimination (compliance range of stimuli: 0.2–2.2 mm/N). Participants were allowed to explore the stimuli unconstrained using their middle finger, ensuring only a single indentation, and the authors observed the use of peak forces around 1 N. Using alarm sounds, the authors constrained the participants' peak forces to within 0.25, 0.49, 0.74 or 0.88 N. The authors observed that for stimuli with rigid surfaces (spring cells) discrimination performance clearly suffered from limiting the peak forces. Performance was also attenuated for stimuli with deformable surfaces, both when the deformable stimuli were actively explored and when they were passively pressed against the participant's finger. Other studies add evidence to the assumption that the use of lower peak forces can result in diminished compliance discrimination for objects with deformable surfaces. Kaim and Drewing (2011) observed for less compliant rubber stimuli (0.15 mm/N), that "low" peak forces (~8 N) resulted in worse discrimination performance than higher forces (~15 N or higher). For more compliant stimuli however (1.5 mm/N), discrimination performance did not vary within the investigated range of peak forces. Nicholson et al. (1998) assessed individual Weber fractions (see Chap. 1) for objects that were even less compliant (0.08 mm/N). Participants considerably differed in their Weber fractions (range 20–6 %) and also in the peak forces that they used to judge compliance (range 80–400 N). Individuals that used lower forces had higher Weber fractions—i.e. compliance discrimination performance was worse. The influence of peak force differences on discrimination performance was smaller the higher the peak forces (Fig. 6.4). Together, the findings from the different studies show that compliance discrimination can benefit from the use of higher peak forces. They also demonstrate that the benefit decreases when peak forces increase (Kaim and Drewing 2011; Nicholson et al. 1998). A comparison across studies (Fig. 6.4) finally suggests that the more compliant the objects are, the less discrimination benefits from the use of higher peak forces.

The suggested relation between peak force, discrimination performance, and object compliance level fits with observations on how participants adjust peak forces to the compliance. Studies regularly show that in unconstrained exploration, humans use higher peak forces the lower the object's compliance (e.g. Freyberger and Färber 2006; Friedman et al. 2008; Fujita and Oyama 1999; Kaim and Drewing 2008). If the peak force that is required in order to achieve good discrimination performance increases with decreasing compliance, the use of higher peak force for less compliant objects represents a useful strategy.

The influences of peak forces and compliance level on discrimination performance can also be well explained by the generated sensory signals. It can be assumed that the deformation of an object comes along with crucial sensory signals for softness judgments. The amount of stimulus deformation caused by an applied force primarily depends on the object's compliance: the less compliant an object is, the higher the force required to effectively deform the surface. As a consequence, in order to discriminate the softness of two less compliant objects, participants have to execute higher forces to produce the required sensory signals—i.e. to effectively and differently deform the two objects—as compared to two more compliant objects. Some

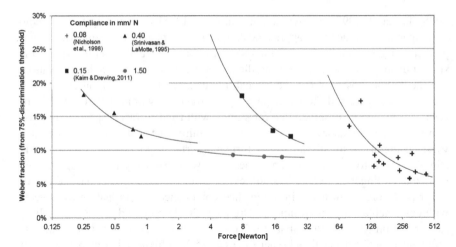

Fig. 6.4 Weber fractions as a function of compliance and peak force based on the data in Kaim and Drewing (2011); Nicholson et al. (1998) and Srinivasan and LaMotte (1995). The *lines* represent the best fit of curvilinear equations of the form Weber Fraction = a + b / Force, suggested by Nicholson et al. (1998). Srinivasan and LaMotte (1995) originally report percentages of correct discriminations between a standard and four comparison stimuli, which were fit to a cumulative Gaussian function using MLE methods (Wichmann and Hill 2001). All Weber fractions were standardized to correspond to 75 % discrimination thresholds. Values represent different experimental conditions (Kaim and Drewing 2011; Srinivasan and LaMotte 1995) or different individuals (Nicholson et al. 1998)

authors (Fujita and Oyama 1999; Nicholson et al. 1998) further linked the sensing of an object's deformation to the area of the deformed finger-stimulus contact region. This area increases with the applied force (Ambrosi et al. 1999; Srinivasan and LaMotte 1995), but the effect is presumably negligible with sufficiently high forces (cf. data in Nicholson et al. 1998). If, indeed, the magnitude of the contact area determines compliance discrimination, it could be expected that compliance discrimination improves with peak force up to a limit. And that the limit is lower for more compliant objects, because a maximal area of contact is achieved with lower force for more compliant objects.

For objects with rigid surfaces, Tan et al. (1995) further showed that participants discriminate compliances using "heuristic" cues that are typically correlated with compliance differences, such as the work executed during exploratory movement (=integral of force over surface displacement), rather than by precisely calculating compliance cues from the ratio between surface displacement and force. Importantly, the value of such correlates depends on the exploratory movements being executed. For example, work increases both with the applied peak force and the compliance of the object, because the surface displacement is larger for more compliant objects. As a consequence, if the compliance of two objects is discriminated using work cues, the differentiation can be successful only if the two objects are explored with similar peak forces, or if the differences in peak force are discounted. Another consequence is that higher peak forces are required to achieve work cues of similar magnitude for objects with low compliance, as compared to objects of high compliance. That is, the use of work cues is able to explain why participants apply and require higher peak forces for the discrimination of less compliant objects with rigid surfaces.

6.3.2 Rate of Force Change

Another movement parameter that has been occasionally discussed in the context of softness perception is the rate of force change (force/time) during object exploration. A literature review shows that the rate of force change and the peak forces that participants aim to achieve are not coupled in a consistent manner. For example, when comparing pressing to tapping movements (Friedman et al. 2008; LaMotte 2000), the higher peak forces used in pressing movements come along with decreased rates of force changes, as compared to the less forceful tapping movements. In contrast, higher peak forces applied to less compliant objects come along with increases or no reliable effect on the rate of force change (LaMotte 2000; compliances 0.19–0.78 mm/N; Friedman et al. 2008; compliances 0.21–7.59 mm/N; deformable objects).

Participants are able to deliberately choose whether they tap or press a stimulus in order to judge its compliance (Friedman et al. 2008; LaMotte 2000), which suggests that they exercise intentional control over the rate of force change. Kaim and Drewing (2008) investigations addressed whether participants intentionally adapt the rate of force change. We measured how rapidly humans approach objects in compliance discrimination (at the moment of finger-stimulus contact) when exploring virtual stimuli modeled with rigid surfaces. Finger velocity can be regarded as a correlate of the rate of force change, if it is measured before the finger deforms the object (cf. LaMotte 2000). We used stimuli of different compliance values (around 4, 10 and 21 mm/N), which were rendered with a force-feedback device. We further manipulated the predictability of the approximate compliance value: Stimuli with different compliance values were presented in different blocks (high predictability) or in random order (low predictability). When participants were able to predict the approximate compliance of the upcoming stimulus, they approached the more compliant stimuli faster than the less compliant ones. When compliance could not be predicted, approach velocity was similar for all compliance values. These results show that participants will adapt their approach velocities (and probably the rate of force change) to the expected compliance of a to-be-explored stimulus. Note that approach velocity has a link to the generation of the so-called "impact tactile cues" to softness (Friedman et al. 2008; Lawrence et al. 2000); cues that are generated during the initial surface contact.

Not much is known about how rate of force change influences discrimination performance. Friedman et al. (2008) reported that free magnitude estimates for the softness of rubber stimuli, with compliances between 0.2 and 7.5 mm/N are similar for individuals that explore with moderately different rates of force change.

6.3.3 Number of Fingers Used

A few studies also investigated how the number of fingers used for exploration affects compliance discrimination. Freyberger and Färber (2006) compared how well humans can discriminate compliances depending on whether they press with the index finger into an object that lies on a fixed support, versus squeezing it between the thumb and index finger while holding it with a pinch grip. They used a number of rubber stimuli of compliances ranging between 0.009 and 2.3 mm/N. Participants judged whether two stimuli felt similarly compliant. The authors report Weber fractions and differential sensitivities (d') for compliance. The results are, however, ambiguous. Whereas differential sensitivities tend to indicate a better discrimination performance for the group of participants that pressed using only the index finger as compared to the group squeezing the stimuli between thumb and index finger, the results in the Weber fractions do not corroborate this interpretation.

Kaim and Drewing (2014) varied the number of fingers that were pressed into compliant objects laying on a fixed support. The objects had deformable surfaces. In the 1-finger-condition, participants were instructed to use only their index finger, in the 3-finger condition they simultaneously used the index, middle, and ring fingers and were instructed to press equally strong with each finger. Interestingly, at the beginning of the experiment, participants reported that the exploration with three finger felt "unfamiliar", suggesting that this exploration strategy was unusual. We measured Weber fractions for lower and higher compliance rubber stimuli (0.16 and 1.3 mm/N, respectively) by combining the method of constant stimuli with a 2-interval forced-choice procedure (N = 8 participants, 196 data points per individual Weber fraction; details as in Kaim and Drewing 2011; Exp. 3). In each trial participants explored a standard and a comparison stimulus, starting with the stimulus on the left side, and decided whether the left or the right stimulus was softer. They were allowed to explore each stimulus only once. There were no further constraints on the exploratory behavior.

Participants adapted their exploratory strategy to the instructed contact mode (1 or 3 fingers), as they did for the compliance level: They applied higher peak forces in the 3-finger condition than in the 1-finger condition, and as observed previously, peak forces were higher for the less compliant as compared to the more compliant stimuli (Fig. 6.5). Due to the larger area of stimulus contact the higher forces in the 3-finger condition still resulted in less object deformation (hard: 3.9 mm, soft: 7.9 mm) compared to the lower applied forces in the 1-finger condition (hard: 5.1 mm, soft: 10.8 mm). Weber fractions were lower, i.e., discrimination performance was better for 3- finger as compared to 1-finger exploration, but only for the more compliant objects. Given that the variation of peak forces in a similar range has been previously shown to have no effect on discrimination performance for the more compliant objects (Kaim and Drewing 2011), this improvement can be explained as being due to a benefit from the additional sensory signals in the 3-finger condition. Overall, the results show that depending on the exact conditions, more fingers can lead to better compliance discrimination.

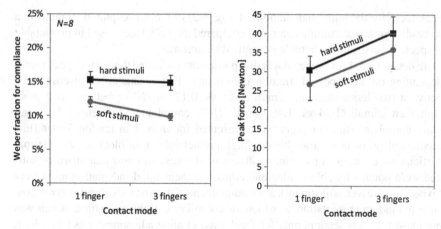

Fig. 6.5 Average Weber fractions (**a**) and average peak forces (**b**) measured while participants explore pairs of more compliant rubber stimuli (*soft*) or pairs of less compliant ones (*hard*) with one or three fingers. Error bars are standard errors. Weber fractions for the more compliant stimuli were lower for the 3-finger condition than for the 1-finger condition, $F(1, 7) = 14.0$, $p < .01$, but they did not significantly vary for the less compliant stimuli; the normal peak force was always higher for the 3-finger condition than for the 1-finger condition (Contact Mode, $F(1, 7) = 68.4$, $p < .001$, Compliance Value, $F(1, 7) = 5.9$, $p < .05$, no interaction)

Taken together, the studies presented in this section suggest that movement parameters (e.g., peak forces, number of fingers) can influence softness perception and that humans sometimes adapt the parameters to the specific stimulus or softness task (e.g., peak force, approach velocity).

6.4 Strategic Adjustment of Exploratory Peak Force

Given that movement parameters appear to influence softness perception and that humans sometimes adjust these parameters to the specific stimuli, Kaim and Drewing (2011) asked whether adjustments of movement parameters are strategically used to improve softness perception. This question leads back to the claim that exploratory movements are executed in a way that optimizes the intake of the relevant sensory information (Gibson 1962; Klatzky and Lederman 1999).

Kaim and Drewing (2011) studied adjustments of peak forces in compliance discrimination and the effects of the applied peak force on discrimination performance in combination, using similar objects and tasks in three experiments. We used silicone rubber stimuli with deformable surfaces that participants explored using a bare finger, pressing their index finger downwards against the surface. The participant's task was always to decide which of the two stimuli felt "softer". The first two experiments focused on expectancy-driven adjustments of peak force, which can be well separated from other influences on exploratory movements. We measured the peak force for the initial indentation (cf. Fig. 6.2) of the first explored stimulus in a to-be-discriminated stimulus pair, and compared the peak forces used in predictable (expected) versus unpredictable perceptual situations.

In detail, Exp. 1 investigated whether participants adjusted their initial peak forces depending on the expected stimulus compliance. Participants either discriminated between two less compliant stimuli (0.15 vs. 0.17 mm/N) or between two more compliant stimuli (1.24 vs. 1.46 mm/N; 15 % compliance difference). Less and more compliant stimulus pairs were presented intermixed in random order (low predictability) or in separate blocks (high predictability). In blocked presentation, participants expected a particular softness of the next stimulus pair (hard or soft) and were potentially able to plan and adjust for their initial indentation movement (expectancy-driven adjustment). Such adjustment of the first movement was not possible in random presentation, in which the compliance of the next stimulus pair was not known. In this session, only feedback-based online-adjustment was possible. It turned out that participants applied higher initial peak forces to the less compliant stimuli than to the more compliant ones (Fig. 6.6, left). Peak forces, however, differed significantly only in the high-predictability conditions, in which participants were able to adjust the peak force already for the first indentation movement. An

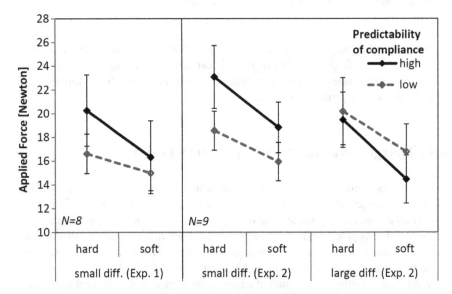

Fig. 6.6 Initial average peak forces while exploring the first object in a compliance discrimination task (Kaim and Drewing 2011; Exp. 1 and 2). Forces were higher for less compliant (*hard*) than for more compliant (*soft*) stimuli, in particular when participants were able to predict the stimulus' approximate compliance. In addition, participants applied lower initial forces in the conditions with large compliance differences than in the conditions with small compliance differences, in the high-predictability conditions

initial force difference was not significant in the low-predictability condition. Participants appeared to strategically adjust their initial exploratory forces to the expected compliance value, when possible.

In Exp. 2, we tested whether participants would also adjust their peak forces to the expected compliance difference between a pair of stimuli. Again, less and more compliant stimulus pairs were presented in low- and high-predictability conditions.In different sessions pairs with small (~15 %) and with large compliance differences were used (~75 %). The results (Fig. 6.6, right), essentially confirm that participants strategically adjust their initial exploratory forces to the expected softness. In addition, participants applied lower initial forces in the conditions with large compliance differences than in the conditions with small compliance differences. This effect was only observed in the high-predictability conditions. In the low-predictability conditions force did not significantly depend on compliance difference. These results suggest that participants also strategically adjust their initial exploratory forces to the expected compliance difference, if they can expect a certain level of compliance.

Experiment 3 assessed how the peak forces chosen in the previous experiments affected discrimination performance. It measured Weber fractions for the less and more compliant stimuli using a 2-interval-forced choice task combined with the method of constant stimuli. The targeted peak forces varied in three conditions: "lower-force", "spontaneous-force", and "higher-force". At the beginning of the experiment, the "spontaneous" peak force adjustments were assessed individually for each participant using the two stimulus pairs with a small compliance difference from Exp. 1. The "lower-force" was then defined as 50 % lower than the spontaneous-force, and the "higher-force" was 50 % higher. Participants were constrained to indent each stimulus once with the prescribed force. For each indentation movement, auditory feedback signaled when participants had achieved the prescribed force.

Measured forces were slightly higher than the prescribed forces (on average by 11 %), but they differed between the force conditions as intended. Figure 6.7 depicts the Weber fractions as a function of the average peak forces in the three force conditions. For the less compliant stimuli, the Weber fractions varied with applied force. They were higher for the lower-force condition than for the other two conditions, but did not significantly differ between the "spontaneous" and the "higher-force" condition. For the more compliant stimuli, there was no distinct relationship between the applied force and the Weber fractions.

Overall, the results of the three experiments show that participants apply a higher peak force if they expect a less compliant object than if they expect a more compliant one.In addition, the applied force is adjusted to account for the compliance difference between the stimuli that are to be discriminated. Participants apply a higher force in order to discriminate between stimuli with a small, as compared to a large, compliance difference, if they expect a specific compliance value. These force adjustments relate to the differential sensitivity for compliance (assessed by the Weber fractions). Participants achieved maximal differential sensitivity for the less compliant stimuli only when applying the spontaneous or higher forces, whereas force did not matter for the more compliant stimuli. This result corroborates a perceptual explanation of the peak force adjustments to stimulus compliance. The participants' high spontaneous

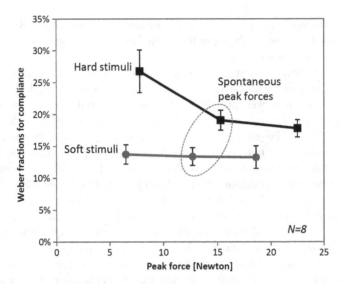

Fig. 6.7 Weber fractions and standard errors as a function of applied peak force (*spontaneous, lower, higher*) and stimulus compliance (*hard* vs. *soft*; Kaim and Drewing 2011; Exp. 3). Note that these data are also part of Fig. 5.4

forces for the less compliant stimuli represent a strategy that almost maximizes differential sensitivity for hard stimuli (relative to other forces), whereas applying high forces to the less compliant stimuli would not result in higher sensitivity, and thus, is not required. Strategically adjusting to compliance differences, however, qualifies this explanation. When expecting large compliance differences, participants lowered their peak forces, which resulted in less than the maximal differential sensitivity for the less compliant stimuli. The reduced forces, however, still allowed them to successfully discriminate the two stimuli of the pair. Kaim and Drewing (2011) estimated the probabilities of correct discriminations from the Weber fractions and found that for the large compliance difference, the probability was always >99 % for each of the three forces. For the small compliance difference and the less compliant stimuli, probabilities of correct responses were 71, 78 and 80 % ("lower-force", "spontaneous", and "higher-force" conditions); for the more compliant stimuli they were 86–87 %. These estimates corroborate the assumption that lower forces suffice for successful discrimination of the large compliance differences (>99 %), whereas they considerably worsen discrimination performance for the low compliance differences and less compliant stimuli. Taken together, these findings suggest that peak forces are indeed strategically adjusted to improve softness perception.

6.5 Conclusion

In order to perceive an object's softness, humans use the specific exploratory movement procedure of *pressure*. This procedure is well suited for the intake of stimulus information that is relevant for softness perception. In addition, the procedure's movement parameters can be strategically adjusted to a particular perceptual situation in order to improve performance.

Such strategic adjustment has at least been show for the parameter 'peak force'. It is an important question for future research which other strategic adjustments are used in softness perception and how they interact. Other movement parameters have been shown to affect performance, such as the type of motion performed (Di Luca et al. 2011) and the number of fingers used in softness perception (Freyberger and Färber 2006), or they have been observed to be adapted by the participants, such as the rate of force change. It is less clear whether participants indeed vary the parameters to maximally improve perception. Moreover, some of these suggested relations between movement parameters and softness perception need better empirical support and only a small section of movement strategies has been studied. Kaim and Drewing (2011) have informally observed that participants prefer to lift the forearm while they explored the stimuli and that some participants hold the finger steeper against the surface of the less compliant stimuli. One may wonder whether such behavior is linked to a need for high exploratory forces for a sufficient stimulus deformation. Likewise, Drewing et al. (2011) observed that a higher number of strokes across a grating stimulus improves discrimination performance for grating parameters. One may wonder whether an analogous relation holds for the number of indentations in softness perception. So, there are still a number of open questions regarding the specific control and adaptation of the EP *pressure*.

Findings from haptic tasks, other than softness discrimination, corroborate the general assumption that participants are able to strategically adjust their movement parameters in order to improve haptic perception: several studies demonstrate that there are mutual dependencies between the ways in which humans explore and the precision and value of the haptic perceptual estimates (Arzamarski et al. 2010; Debats et al. 2010; Drewing 2012). Some studies directly demonstrate the link between spontaneously preferred exploratory movement parameters and good perceptual performance (Arzamarski et al. 2010; Drewing 2012; Drewing and Kaim 2009; Gamzu and Ahissar 2001). Gamzu and Ahissar (2001), for example, observed that during spontaneous exploration of haptic gratings, some of their participants reduced exploratory velocity after a few trials, and that the reduced velocity allowed for better discrimination performance than the initial higher velocity. Our own studies demonstrated that the exploratory directions relative to the body affect discrimination performance for small virtual bumps, and that participants preferred to explore along directions that were associated with superior discrimination performance (Drewing and Kaim 2009). In addition, if the exploratory direction that yields optimal performance was manipulated, participants changed their strategic preferences accordingly (Drewing 2012). It is noteworthy that all studies that demonstrate successful exploratory adjustments

(Arzamarski et al. 2010; Drewing 2012; Gamzu and Ahissar 2001) also show that these adjustments require sufficient experience with the perceptual situation.

Optimal adjustments might sometimes represent a costly strategy. In the case of the high peak forces used to improve softness discrimination, the study of Kaim and Drewing (2011) showed that this exhausting strategy is only used if a high level of differential sensitivity is required. Strategic adaptations of exploratory forces aimed for high levels of performance in the discrimination task, rather than for perceptual optimization. Similarly, in the study showing that a higher number of strokes improved grating discrimination up to a maximum performance for 6–7 strokes (Drewing et al. 2011), the spontaneously preferred number of strokes was slightly below 6, resulting in close-to-optimal performance while keeping motor costs low. A recent study (Saig et al. 2012) investigated a simplified haptic task, in which participants discriminated object positions using a whisker in either hand, and presents a model on how sensory information is accumulated over time while motor costs are minimized. An interesting question for future research is how motor costs and perceptual benefits of haptic optimization strategies are balanced in natural haptic tasks.

Acknowledgments I wish to thank two anonymous reviewers and Alexandra Lezkan for their helpful criticisms, suggestions, and comments on an earlier draft, and Steven A. Cholewiak for native-speaker advice.

References

Ambrosi G, Bicchi A, Rossi DD, Scilingo EP (1999) The role of contact area spread rate in haptic discrimination of softness. In: Proceedings IEEE international conference on robotics and automation, pp 305–310

Arzamarski R, Isenhower RW, Kay BA, Turvey MT, Michaels CF (2010) Effects of intention and learning on attention to information in dynamic touch. Atten Percept Psychophys 72:721–735

Bergmann Tiest WM, Kappers AML (2008) Kinesthetic and cutaneous contributions to the perception of compressibility. In: Ferre M (ed) Haptics: perception, devices and scenarios. Lecture notes in computer science, vol 5024. Springer, Heidelberg, pp 255–264

Bicchi A, Scilingo E, Dente D, Sgambelluri N (2005) Tactile flow and haptic discrimination of softness in multi-point interaction with real and virtual objects. In: Barbagli F, Prattichizzo D, Salisbury K (eds) Series STAR: Springer tracts in advanced robotics, vol 18. Springer, Heidelberg, pp 165–176

Choi S, Walker L, Tan HZ, Crittenden S, Reifenberger R (2005) Force constancy and its effect on haptic perception of virtual surfaces. ACM Trans Appl Percept 2:89–105

Debats NB, van de Langenberg RW, Kingma I, Smeets JBJ, Beek PB (2010) Exploratory movements determine cue weighting in haptic length perception of handheld rods. J Neurophysiol 104:2821–2830

Di Luca M, Knorlein B, Ernst MO, Harders M (2011) Effects of visual-haptic asynchronies and loading-unloading movements on compliance perception. Brain Res Bull 85(5):245–259

Drewing K, Kaim L (2009) Haptic shape perception from force and position signals varies with exploratory movement direction and the exploring finger. Atten Percept Psychophys 71:1174–1184

Drewing K (2012) After experience with the task humans actively optimize shape discrimination in touch by utilizing effects of exploratory movement direction. Acta Psychol 141:295–303

Drewing K, Lezkan A, Ludwig S (2011) Texture discrimination in active touch: effects of the extension of the exploration and their exploitation. In: Basodogan C, Choi S, Harders M, Jones L, Yokokohji Y (eds) Conference proceedings: IEEE World haptics conference 2011, The Institute of Electrical and Electronics Engineers (IEEE) Catalog number CFP11365-USB, pp 215–220

Freyberger FKB, Färber B (2006) Compliance discrimination of deformable objects by squeezing with one and two fingers. Proceedings of eurohaptics, Paris, France, pp 271–276

Friedman RM, Hester KD, Green BG, LaMotte RH (2008) Magnitude estimation of softness. Exp Brain Res 191:133–142

Fujita K, Oyama Y (1999) Control strategies in human pinch motion to detect the hardness of an object. IEEE SMC '99 Conference proceedings, vol 2

Gamzu E, Ahissar E (2001) Importance of temporal cues for tactile spatial-frequency discrimination. J Neurosci 21:7416–7427

Gibson JJ (1962) Observations on active touch. Psychol Rev 69:477–490

Kaim L, Drewing K (2008) Exploratory movement parameters vary with stimulus stiffness. In: Ferre M (ed) Haptics: perception, devices and scenarios. Lecture notes in computer science, vol 5024. Springer: Heidelberg, pp 313–318

Kaim L, Drewing K (2011) Exploratory strategies in haptic softness discrimination are tuned to achieve high levels of task performance. IEEE Trans Haptics 4:242–252

Kaim L, Drewing K (2014) Haptic softness discrimination of deformable objects by pressing with one or three fingers (unpublished raw data)

Klatzky RL, Lederman SJ (1999) The haptic glance: a route to rapid object identification and manipulation. In: Gopher D, Koriat A (eds) Attention and performance XVII: cognitive regulation of performance: interaction of theory and application. Erlbaum, Mahwah, pp 165–196

Klatzky RL, Lederman SJ, Reed CL (1989) Haptic integration of object properties: texture, hardness, and planar contour. J Exp Psychol Hum Percept Perform 15:45–57

LaMotte RH (2000) Softness discrimination with a tool. J Neurophysiol 83:1777–1786

Lawrence DA, Pao LY, Dougherty AM, Salada MA, Pavlou Y (2000) Rate-hardness: a new performance metric for haptic interfaces. IEEE Trans Robot Autom 16:357–371

Lederman SJ, Klatzky RL (1987) Hand movements: a window into haptic object recognition. Cogn Psychol 19:342–368

Lederman SJ, Klatzky RL (1990) Haptic classification of common objects: knowledge-driven exploration. Cogn Psychol 22:421–459

Lederman SJ, Klatzky RL (1993) Extracting object properties through haptic exploration. Acta Psychol 84:29–40

Nefs HT, Kappers AML, Koenderink JJ (2002) Frequency discrimination between and within line gratings by dynamic touch. Percept Psychophys 64:969–980

Nicholson LL, Maher CG, Adams RD (1998) Hand contact area, force applied and early non-linear stiffness (toe) in manual stiffness discrimination task. Manual Therapy 3:212–219

Riley MA, Wagman JB, Santana M, Carello C, Turvey MT (2002) Perceptual behavior: recurrence analysis of a haptic exploratory procedure. Perception 31:481–510

Saig A, Gordon G, Assa E, Arieli A, Ahissar E (2012) Motor-sensory confluence in tactile perception. J Neurosci 32(40):14022–14032

Srinivasan MA, LaMotte RH (1995) Tactual discrimination of softness. J Neurophysiol 73:88–101

Tan HZ, Durlach NI, Beauregard GL, Srinivasan MA (1995) Manual discrimination of compliance using active pinch grasp: the role of force and work and cues. Percept Psychophys 57:495–510

Tan HZ, Durlach NI, Shao Y, Wei M (1993) Manual resolution of compliance when work and force cues are minimized. In: Kazerooni H, Colgate JE, Adelstein B (eds) Advances in robotics, mechatronics, and haptic interfaces, vol 49. ASME, New Orleans, pp 99–104

[Wichmann and HillWichmann and Hill2001]bib33 Wichmann FA, Hill NJ (2001) The psychometric function: I. Fitting, sampling and goodness-of-fit. Percept Psychophys 63: 1293–1313

Chapter 7
The Perception of the Centre of Elastic Force Fields: A Model of Integration of the Force and Position Signals

Gabriel Baud-Bovy

7.1 Introduction

As was noted by Charles Bell in 1833, softness and hardness perception relies on the capacity of the haptic system to simultaneously produce and sense force and movement:

> "Without a sense of muscular action or a consciousness of the degree of effort made, the proper sense of touch could hardly be an inlet to knowledge at all... The property of the hand of ascertaining the distance, the size, the weight, the form, the hardness and softness, the roughness or smoothness of objects results from the combined perception—through the sensibility of the proper organ of touch and motion of the arm, hand and fingers." Charles Bell (1833) *Bridgewater Treatise on the Hand* (I, 2), 193,202–203.

While scholars of the sense of touch have always been aware of the importance of movement in this sensory modality (e.g., Sherrington 1900; Gibson 1962), the experimental study of this contribution started in earnest much later. Klatzky and Lederman conducted a series of landmark studies during the 80s and 90s about the manner in which people recognize objects and their physical attributes haptically (Lederman and Klatsky 1987). They found that people used specialized hand movements—which they called exploratory procedures—to extract different types of information from the objects (e.g. shape, weight or softness). They also found that people could identify objects very quickly so long as their hand movements were unrestricted (Klatzky and Lederman 1995). While early studies on the sensitivity of the haptic system focused on vibro-tactile stimuli that were applied on the skin and passively experienced, researchers have since started to measure the sensory thresholds for physical attributes, which require an active movement to be perceived such

G. Baud-Bovy (✉)
Department of Robotics, Brain and Cognitive Sciences (RBCS), Istituto Italiano di Tecnologia, Via Morego 30, 16123 Genoa, Italy
e-mail: baud-bovy.gabriel@iit.it

G. Baud-Bovy
Faculty of Psychology, San Raffaele Vita-Salute University, Milan, Italy

© Springer-Verlag London 2014
M. Di Luca (ed.), *Multisensory Softness*, Springer Series on Touch and Haptic Systems,
DOI 10.1007/978-1-4471-6533-0_7

as stiffness (e.g., Tan et al. 1995), viscosity (e.g., Beauregard et al. 1995; Jones and Hunter 1993) and mass (e.g., Baud-Bovy and Schochia 2009; Brodie and Ross 1984; Ross and Brodie 1987).

At the physical level, the resistance to deformation of soft materials is often modeled by a spring that produces an elastic force when one pushes against the surface. In this simple model, the main parameter that characterizes the deformable object is its stiffness, which expresses the force generated by the spring as a function of the depth of the penetration. In a seminal study of stiffness perception, Tan et al. (1995) have shown that the discrimination threshold for stiffness perception is contingent on whether the same or different movements are used to feel the stiffness (see also Chap. 6). More recent studies of softness perception have shown that softness perception, like that of stiffness, is influenced not only by the modality of exploration of the force field, but also by other sensory modalities (see Chaps. 2–4).

Models of stiffness perception might be classified according to the type of information provided by the sensory apparatus and the manner these signals are integrated. In some models of softness perception, stiffness is directly sensed by the peripheral apparatus. For example, when touching deformable objects, the rate of change of average pressure might be used to estimate stiffness because it is invariant with respect to indentation velocity (Srinivasan and LaMotte 1995). These authors have proposed that this variable might be directly encoded in the population response of SAIs (Srinivisan and LaMotte 1996).

In other models of stiffness perception, separate position and force signals must be integrated to yield stiffness. Some of these models assume that the time-derivatives of these signals is available. For example, when touching compliant objects with a rigid probe, Srinivasan and LaMotte (1995) have proposed that stiffness might be estimated from information about the rate of force and indentation velocity provided by tactile and kinesthetic inputs respectively. Other models do not assume that the time-derivatives are available. Of these, some posit that perceived stiffness corresponds to the ratio between maximum interaction force and perceived penetration (Pressman et al. 2007, 2011). Other assume that the whole trajectory is taken into consideration and that stiffness is obtained by regressing position on force or the opposite depending on whether they cross the boundary (Nisky et al. 2008). Finally, it has also been proposed that the information acquired during the loading and unloading phases should be weighted differently (Di Luca et al. 2011). While all these models assume that positional and force information are combined in stiffness perception, there is no consensus on the manner in which such integration occurs (see Chap. 9 for a detailed discussion of these models).

This chapter investigates the accuracy and precision with which the centre of an elastic force field can be identified. In this study, the force field was rendered in the lateral direction by a haptic device (see Fig. 7.1). To report their answer, participants were told to bring their hand to the position where the force felt the smallest and/or changed direction (both formulations were used to explain the task). The task becomes challenging when the stiffness of the force field is small. In this case, participants cannot perceive any force in an extended region around the centre of the force field because the force is below their perceptual threshold.

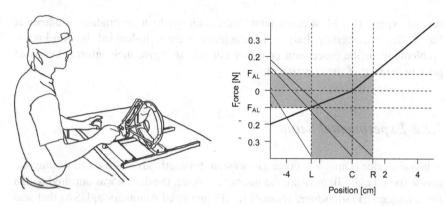

Fig. 7.1 *Left* The experimental setup. *Right* Outwardly oriented asymmetric elastic force field with a stiffness of $k_L = 4$ [N/m] on the *left side* and $k_R = 8$ [N/m] on the *right side*. The shaded area indicates the region defined by the limits L and R around the central position C where the force is not perceived because it is below the force absolute threshold (F_{AL}). The total workspace extended over 20 cm and the maximum force was limited to 1 N (only the central part of the workspace is represented) [adapted with permission from Bocca and Baud-Bovy (2009)]

The main contribution of this chapter is to propose a model of how force and positional information might be integrated in this task. A priori, estimating the stiffness of the force field may help in identifying its central position because stiffness combined with the force perceived at the current position could be used to infer the distance between the current and central positions. Alternatively, this task could be performed by simply sensing the position where the force becomes imperceptible or changes direction.

This chapter is organized as follows. Section 7.2 describes the experimental apparatus and task in more detail and Sect. 7.3 presents briefly the results of two previous experiments, reported in Bocca and Baud-Bovy (2009). Section 7.4 describes a model of how the participants might estimate the central position of an elastic force field from the information they can sense while they try to identify the central position. The predictive accuracy of the model parameters and predictions of the model are fully worked out. Unlike the model described in Bocca and Baud-Bovy (2009), this model is able to predict both the systematic and variable error patterns observed in the two experiments. The assumptions and limitations of this model are discussed in Sect. 7.5.

7.2 Methods

7.2.1 Participants

14 participants (4 males) aged 19–38 years (mean 23.7) took part in the first experiment. 30 participants (12 males) aged 19–50 (mean 24.46) took part in the

second experiment. Most participants were undergraduate or graduate students at San Raffaele University, Italy. All participants were right-handed, had no known problems of haptic perception or motor control, and gave their informed consent prior to the start of the experiment.

7.2.2 Experimental Setup

A haptic device (Omega.3, Force Dimension, Switzerland) was used to display the elastic force fields. To measure the interaction force, the device was outfitted with a force-torque (FT) transducer (Nano 17, ATI Industrial Automation, USA) that was mounted between the custom-made triangular plate connecting the three parallel robot arms and the spherical handle (diameter ∼1.5 cm) grasped by the participant.

The haptic device was controlled by a program running on an AMD Sempron Processor computer with a Windows XP operating system. The computer was equipped with a 16-bit PCI DAQ card (6034 E, National Instruments) to acquire and digitize signals from the FT sensor. The C/C++ program controlling the device used Force Dimension DHD API to control the haptic device and National Instrument NIDAQmx API to control the DAQ card.

The device produced a strong visco-elastic force field (stiffness = 1,200 N/m, damping = 50 N/m/s) that constrained the movements of the end-effector along a horizontal laterally oriented line. The device displayed the elastic force field along the line direction where the end-effector movements were not constrained. The length of the workspace along this direction was 20 cm.

To improve the force-rendering performance of the device, the program implemented a closed-loop force control law: '

$$f_{cmd} = f_d + k_f(f_s - f_d) \qquad (7.1)$$

where f_{cmd} is the force command sent to the device, f_d is the desired force that corresponds to the elastic force field, f_s is the interaction force measured by the force/torque sensor and $k_f = 7.5$ is the gain of the force-error feedback. The program polled the handle position (x) and the actual force (f_s), and computed the force command (f_{cmd}) at 1 kHz.

In Experiment 1, the sensed force was passed through an exponential filter (time constant = 0.002 s). Adjustments were made in Experiment 2 to the control law to avoid instabilities that sometimes occurred at the limit of the workspace. In particular, to reduce sensory noise, the force was sampled continuously at 33 kHz and the last 33 samples were used haptic loop to estimate the current force [see Baud-Bovy and Gatti (2010a, b), for more details].

Force Dimension DHD antigravity compensation scheme was activated throughout the experiment (end-effector mass parameter = 0.08 kg). The current time, end-effector position, desired and sensed forces were saved at 50 Hz for off-line analyses.

7.2.3 Experimental Procedure and Stimuli

Participants grasped the spherical handle of the device with the thumb and index finger of the right hand using a key grasp. The height of the seat was adjusted so that the elbow was flexed at 90° (Fig. 7.1). The centre of the device workspace was aligned with the body midline in the first experiment and with the shoulder in the second experiment because the position felt more comfortable.

The *stimulus* consisted of a one-dimensional elastic force field that pushed the handle laterally, away from a central position:

$$f_d = \begin{cases} k_L(x - x_c) & \text{if } x \le x_c \\ k_R(x - x_c) & \text{otherwise} \end{cases} \tag{7.2}$$

where f_d is the desired force, and k_L and k_R are the stiffnesses on the left and right side of the central position x_c respectively. The force was oriented outwardly rather than inwardly to avoid the possibility of the force moving the participant's hand toward the centre (this possibility is in any case remote given the fact that the force is too low— typically less than 0.5 — put the participant's hand/arm in motion). Participants were told to maintain a firm grasp throughout the trial. The force magnitude was limited to ±1 N.

A brief familiarization period was offered at the beginning of the experiment to teach the structure of the trial to the participant. Each trial started with a 4 s long positioning period during which the device brought the hand to a starting position. Then, the computer produced a "beep" indicating that the subject could start to explore the force field by moving the end-effector laterally along the unconstrained direction. Participants were instructed to find the point along the line where they did not feel any force and to press a button held in their left hand once they had found it. They were also told that the force changed direction at this point. The exploration time was not constrained but the experimenter encouraged the subject to respond faster if he or she took more than 15–20 s to respond. At the end of the trial, the experimenter had the option to repeat the trial (which was exercised in a few instances where the participant did not perform as expected).

In *Experiment 1*, the force field was always symmetric, i.e. the stiffness on either side of the central position was equal ($k = k_L = k_R$). Each possible stiffness value ($k = 4, 8, 16, 32,$ or 64 N/m) was presented once with a different combination of starting and central positions. There were 12 such combinations, each obtained by selecting one of four possible positions placed at ±1.5 or ±4.5 cm relative to the workspace as the central position and one of the remaining three positions as starting position. Altogether, the experiment included 60 trials.

In *Experiment 2*, the stiffness of the force field could be different on the left and right sides of the central position. The force field was obtained by combining three stiffness values (i.e. 4, 8 and 16 N/m) pairwise, thus obtaining six asymmetric force fields and three symmetric ones. To vary the central position of the force field, we randomly generated a central position inside three different regions (left region: from

−3.5 to −1.5 cm, centre region: from −1 to 1 cm, and right region: from −1.5 to 3.5 cm). The two starting positions were spaced on either side, at an equal but random distance which varied between 1 and 3 cm. The total number of trials per subject was 54 (9 stiffness conditions ×3 regions ×2 starting positions, no replication).

In the two experiments, the order of presentation of the stimuli was randomized for each participant and the duration of the experimental session never exceeded 40 min.

7.2.4 Data Analysis

The positioning error was the main performance measure. For each stiffness condition and subject, we computed the average and standard deviation of the positioning error between the participant's response and the actual centre of the force field. Then, we computed the between-subject and within-subject variable errors for each experimental condition. The between-subject variable error corresponded to the standard deviation of the average positioning errors for each subject while the within-subject variable error corresponded to the across-subject average of the standard deviations of the positioning error computed for each subject and condition separately. In both experiments, we identified a few outlier responses and replaced them by the corresponding condition average. In all, less than 2 % of the trials were considered outliers [see Bocca and Baud-Bovy (2009) for a more detailed description].

Before presenting the results of the experiments, it is important to say a few words about the use of haptic devices in perceptual studies. While simulating an elastic force field with a haptic device is in principle very simple since the desired force is proportional to the position, it can be challenging to render a force with high accuracy in practice because the device is not perfectly "transparent" (Lawrence 1993; Parietti et al. 2011). The problem is that haptic devices are mechanical devices and, as such, endowed with physical properties (e.g. mass, Coulomb friction, etc.) that introduce a difference between the desired force set in the program and the actual force rendered by the device (Salisbury et al. 2009). In order to address this issue, a force sensor was mounted near the handle that was grasped by the participant to measure the actual force. Then this information was used in a quick feedback loop to render the force as described previously (see Sect. 7.2.2).

To assess the quality of the stimulus in this study [for a more detailed analysis of the setup, see Gurari and Baud-Bovy (2014)], we computed for each trial the average, standard deviation and root mean square (RMS) of the force error, i.e., the difference between the desired and measured force ($f_e = f_s - f_d$). We also identified the *zero-force points*, i.e. the points in the scanning movement where the force measured by the sensor was zero (Fig. 7.2, inset). For each trial, we computed the average and standard deviation of the zero-force points.

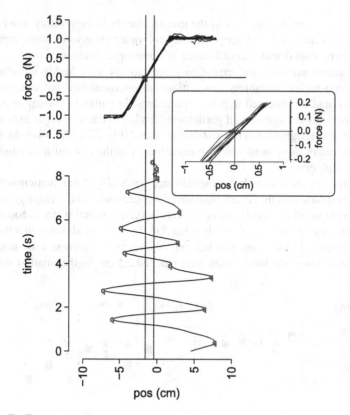

Fig. 7.2 *Top* Force measured during the trial. *Inset* Zoom of the measured force around the central position. The standard deviation of position of the zero-force points (*small empty circles*) is 2 mm in this trial. *Bottom* Scanning movements during a trial (stiffness $k = 32$ N/m). The force field central position and the trajectory final position are denoted by solid and dashed lines respectively [adapted with permission from Bocca and Baud-Bovy (2009)]

7.3 Results

The direction of the force field was clearly identifiable away from the centre, where it reached a maximum value of 1 N (see Fig. 7.2). As participants moved the handle closer to the centre, the intensity of this sensation decreased until it became imperceptible. For the force fields used in this study, the region of uncertainty about the direction of the force could extend over several centimeters.

As expected, participants moved their hand sideways to identify the centre of the elastic force field in the two experiments. In Experiment 1, the duration of the exploration period varied from 6 to 28 s (5 and 95 % quantiles, median $= 12$ s). During the exploration, participants produced on average 15 ± 3.5 (average $\pm$ SD) movements. The average length of these exploratory movements was 5.1 ± 4.6 cm. In experiment 2, the median exploration time was 15.0 s, the average number of exploratory movements was 19, and the average length of exploratory movements was 3.5 cm. At this level of description, the exploration trajectories were similar in the two experiments despite the difference in symmetry/asymmetry.

At the individual level, the exploratory movements were highly idiosyncratic (high between-subject variability, low within-subject variability) but some general tendencies could be observed in both experiments. In particular, visual inspection of the trajectories showed that all participants tended to execute larger exploratory movements at the beginning of the trial than at its end (Fig. 7.2). Moreover, the length of the exploratory movement was in general longer in the low stiffness conditions than in the high ones.

In the first experiment with symmetric force fields (Fig. 7.3a), participants centred their response near the central position of the force field. The average systematic error was very small in all stiffness conditions (<0.2 cm for all stiffness condition). The response variability was inversely related to the stiffness showing that the centre of the force field was identified less reliably when the stiffness was small (the standard deviation of the localization error reached 2.5 cm for the smallest stiffness

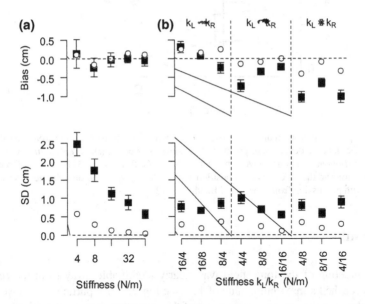

Fig. 7.3 Bias and variability of the estimated force field centres. *Top panels* Average positioning error (*solid squares*) and average position of the zero-force points (*empty circles*) for each stiffness condition. *Vertical bars* denote standard error of the mean. *Bottom panels* Pooled estimate of the standard deviation of the positioning error (*solid squares*) and the average value of the within-trial standard deviation of zero-force points (*empty circles*) for each stiffness condition. **a** Symmetric force field (Exp. 1). Note the log scale of the X axis. **b** Asymmetric force field (Exp. 2). In the *top panel*, note the presence of a bias toward the weak side: toward the right (*positive value*) when $k_R < k_L$ and toward the left (*negative value*) when $k_L < k_R$ [adapted with permission from Bocca and Baud-Bovy (2009)]

condition). A repeated-measure ANOVA confirmed that the stiffness factor had a statistically significant effect on the within-subject standard deviation of the position error ($F(4, 52) = 24.702$, $P < 0.001$).

The main finding of the second experiment was the observation of a systematic bias toward the side of the weakest force field (Fig. 7.3b, top panel). In addition, we found a general bias toward the left (-0.4 ± 1.2 cm) that increased when the stiffness was small. The variable error was smaller in the second experiment relative to the first experiment. The reason for this discrepancy is not clear but it is possible that the improvement in the control law of the haptic device in the second experiment might have contributed to a decrease in the response variability at low stiffness. That being said, the error pattern in the second experiment was qualitatively similar to the one observed in the first experiment. In both experiments, the response variability increased when stiffness decreased but did not tend toward zero when the stiffness increased (Fig. 7.3, bottom panels).

For both experiments, the position of the force field's centre and the hand position was randomized at the beginning of each trial. In both experiments, we found a bias toward the centre of the workspace (range effect) and a weak bias toward the starting position. A more detailed analysis of these effects can be found in Bocca and Baud-Bovy (2009).

7.3.1 Force Rendering Accuracy

The main measure of the rendering performance is the RMS of force error. In the first experiment, the average RMS force error was 0.027 N. The force error never exceeded 0.03 N and was on average well centreed. The mean force error was 0.002 N and the average of the within-trial standard deviation was 0.025 N. In the second experiment, the average RMS force error was 0.021 N which indicates that the changes in the control law in the second experiment slightly improved the quality of the stimulus. In any case, these results constitute a 10-fold improvement of similar measures obtained with an open-loop control of the haptic interface (data not shown).

Although the force error was small, it could not be completely ignored. At low stiffness in particular, the variation of the measured force during the exploratory movement lead to a dispersion of the points where the force was zero (Fig. 7.2, inset panel). In the first experiment, the zero-force points were near the central position (average systematic error was -0.2 mm). Their dispersion (SD) increased from 0.4 mm in the stiffest condition to 5.8 mm in the least stiff condition (see Fig. 7.3). In the second experiment, the dispersion of the zero-force points was 4.5 mm when $k_R = k_L = 4$ N/m, which constitutes a small improvement relative to the first experiment. Note that the variability of the zero-points approximately matches the variability of the force error divided by the stiffness (see also Fig. 7.5).

7.4 The Bisection Model

In this section, the model is progressively built up, starting from the basic assumptions that the participants cannot sense force below some minimal value. We first present the model in relation to the results of the first experiment. The same model is then used to account for the results of second experiment. In both cases, the model is expected to predict the systematic and variable errors of both experiments.

7.4.1 Initial Assumptions

According to Hooke's Law, the force in a linear elastic field is

$$F = k\,\Delta x \tag{7.3}$$

where Δx is the distance from the centre of the force field and k is the stiffness (see Chap. 1). Assuming that there is a threshold below which a person does not perceive any force, then no force will be perceived in the neighborhood of the force field centre. The limits of this interval are

$$L = -\frac{F_{AL}}{k_L} \qquad R = \frac{F_{AL}}{k_R} \tag{7.4}$$

where F_{AL} is the force threshold and k_R and k_L denote the respective stiffness of the force field at the two sides. Note that the size of the interval can always be increased experimentally by decreasing the stiffness if the force threshold differs from zero ($F_{AL} > 0$).

As the observer is unable to perceive any force in this interval, he or she must estimate the central position of the force field from cues coming from outside this interval. In other words, the observer must explore the force field by moving their hand and integrating the information acquired during this exploration to estimate the central position of the force field.

The bisection model presented here assumes that the observer detects the two limits of the interval and identifies a middle position as the central the position of the force field. Mathematically, the model posits that the response corresponds to the weighted average between the left and right limits:

$$X_w = (1 - w)\,L + wR \tag{7.5}$$

where $0 \le w \le 1$ is the weight of the right limit. A true bisection is obtained when the two limits are weighted equally ($w = 0.5$). Values of w deviating from 0.5 reflect a systematic bias toward one or the other limit.

7.4.2 Sensory Noise and Response Variability

In Psychophysics, the *absolute threshold* is operationally defined as the intensity of the stimulus that is detected in 50 % of the presentations. As implied by this definition, any model of the process yielding the response must be formulated in probabilistic terms to account for the variability of the response. In fact, a primary assumption of classical threshold theory is that the threshold varies over time (Gescheider 1997, p. 75). Moreover, it is customary to model the threshold as a normally-distributed random variable.[1] In a detection experiment, the mean of this random variable corresponds to the absolute threshold, while the variance indicates the degree of fluctuations of the threshold due to uncontrolled experimental, biological and/or psychological factors.

In the context of this study, the sensory threshold F_{AL} is represented by a random variable with expected value $\mu_F > 0$ and standard deviation σ_F. The variance σ_x^2 of the response can be obtained by substituting the definitions of the limits (Eq. 7.4) in the model of the response (Eq. 7.5) and by applying the mathematical properties of the variance:

$$\sigma_X^2 = var\,[X_w] = var\,[(1-w)L + wR] = (1-w)^2 var\,[L] + w^2 var\,[R]$$
$$= \left(\frac{1-w}{k_L}\right)^2 var\,[F_{AL}] + \left(\frac{w}{k_R}\right)^2 var\,[F_{AL}] = \sigma_F^2 \left\{ \left(\frac{1-w}{k_L}\right)^2 + \left(\frac{w}{k_R}\right)^2 \right\}$$

$$(7.6)$$

The estimated positions of the two limits, L and R, of the interval below the perceptual threshold are assumed to be uncorrelated. This assumption is reasonable because the two limits are assessed at different times during the exploration of the force field.

When the force field is symmetric ($k_R = k_L$), this equation becomes

$$\sigma_X^2 = (2w(w-1) + 1)\,\frac{\sigma_F^2}{k^2} = W^2 \frac{\sigma_F^2}{k^2}. \qquad (7.7)$$

where the factor $W = \sqrt{2w(w-1)+1}$ depends only on the weight w. Note that this model predicts an inverse and proportional relationship between the stiffness and the standard deviation of the response since, by definition, the variability of the force threshold is constant and the weight w is also assumed to be fixed:

$$W\sigma_F = k\sigma_X \equiv \text{constant}. \qquad (7.8)$$

---

[1] An alternative view is to conceptualize the processes yielding a sensation as intrinsically noisy (Thurstone 1927). In this case, the sensation elicited by the stimulus is modeled as a random variable while the threshold is assumed to be fixed. Moreover, the response variability arises from the fluctuations of the sensation instead of the threshold. The two threshold models are not distinguishable empirically without making additional assumptions. Depending on the view, the response variability can therefore be interpreted either as quantifying the variability of the threshold or the amount of noise in the sensory channel (Gescheider 1997; Macmillan and Creelman 2005).

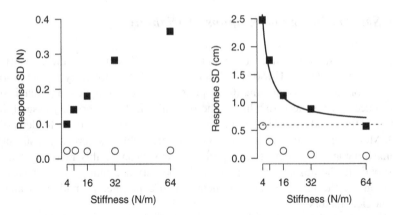

Fig. 7.4 Modeling of the response variability in Experiment 1. The *solid squares* indicate the average within-subject standard deviations of the localization errors while the *empty circles* represent the standard deviations of the zero-force points. *Left* Both standard deviation are expressed in Newton by scaling the positional standard deviation with the corresponding stiffness. Note the increase as a function of the stiffness. *Right* The *solid line* represents the prediction of the model as a function of the stiffness (see Eq. 7.11) while the dashed horizontal line represents the estimated value of the positional noise $\sqrt{0.5}\sigma_P$ (i.e., the intercept of Eq. 7.11). Note that the *panel* shows the same data as in the *bottom panel* of Fig. 7.3a with a linear scale on the X axis

To test this prediction, we multiplied the standard deviation of the response by the stiffness to transform it in force units. Figure 7.4 (*left panel*) shows that the variability of the positioning error expressed in Newton increases as a function of the stiffness, contrary to what is predicted by the model (Eq. 7.8) since W is constant. A repeated-measure ANOVA confirmed that the stiffness had an effect on the variability of the response expressed in Newton ($F(4, 52) = 11.03$, $P < 0.001$). This variability increase cannot be explained by a decrease in the force rendering quality for higher stiffness. As a matter of fact, the standard deviation of the zero-force points multiplied by the stiffness remained constant once expressed in Newton. The average standard deviation of the zero-force points expressed in Newton was 0.023 N, which more or less corresponds to standard deviation of the force error reported in Sect. 7.3.1 (0.025 N).

This negative result clearly implies that something is wrong or lacking in the initial model. In fact, going back to the definition of the model, another of its implications is that the force field centre should be localized without variable error when the stiffness is infinite (force threshold variability is divided by the stiffness in Eqs. 7.6 and 7.7). In other words, the model assumes a noiseless position sense. However, the plot of the response variability as a function of the stiffness clearly suggests the presence of an asymptotic floor value for the variability of the response (see Fig. 7.3a). The origin of this flaw in the model is that the variability of the response is entirely attributed to the variability of the force threshold, which is then propagated according to the mathematical specification of the model.

A more realistic formulation of the model is to assume a second source of error with respect to the perceived position of the force threshold. This can be modeled mathematically by introducing a second random variable P with a mean equal to zero and standard deviation σ_P in the definition of the perceived limit of the interval without force

$$L = -\frac{F_{AL}}{k_L} + P \qquad R = \frac{F_{AL}}{k_R} + P \qquad (7.9)$$

The random variable P represents an additional source of noise and uncertainty related to the position in space of the point where the force threshold is reached. The variance of the responses predicted by the model is now

$$\sigma_X^2 = (1 - w)^2 \, var \, [L] + w^2 var \, [R]$$
$$= (1 - w)^2 \, var \left[-\frac{F_{AL}}{k_L} + P \right] + w^2 var \left[\frac{F_{AL}}{k_R} + P \right]$$
$$= (1 - w)^2 \left(\frac{\sigma_F^2}{k_L^2} + \sigma_P^2 \right) + w^2 \left(\frac{\sigma_F^2}{k_R^2} + \sigma_P^2 \right) \qquad (7.10)$$

If one assumes that the estimated position of the two limits, L and R, are uncorrelated as before, and that the two random variables F_{AL} and P are also uncorrelated. This latter assumption makes sense because these two random variables model noise in two different sensory channels.

When the force field is symmetric, the standard deviation of the responses corresponds to

$$\sigma_X = W \left(\frac{\sigma_F}{k} + \sigma_P \right) \qquad (7.11)$$

where W is defined as in Eqs. 7.7 and 7.8. In other words, the model predicts that the standard deviation of the responses divided by W is linearly related to the inverse of stiffness and bottoms up at some asymptotic value. The intercept and slope of this relationship can be interpreted as the standard deviations of P and F_{AL} respectively. Assuming that both limits are weighted equally ($w = 0.5$ or $W = \sqrt{0.5}$) and regressing the standard deviation of the responses observed in the first experiment divided by W on the inverse of the stiffness yielded estimates for the force threshold standard deviation of 0.11 N and 8.4 mm for the positional precision. Figure 7.4 (right panel) shows that this model fits the data well.

7.4.3 Force Threshold and Response Bias

The systematic error or bias predicted by the model corresponds to the expected value of the response (Eq. 7.5):

$$\mu_X = E[X_w] = E\left[(1-w)\frac{-F_{AL}}{k_L} + w\frac{F_{AL}}{k_R} + 2P\right]$$

$$= \mu_F\left(-\frac{(1-w)}{k_L} + \frac{w}{k_R}\right) \tag{7.12}$$

Interestingly, the bias depends on the force threshold expected value μ_F. In other words, it is in principle possible to estimate the force threshold from the systematic errors.

One caveat is that estimating this parameter is problematic in the first experiment because the force field is symmetric ($k_L = k_R$). In this case, Eq. 7.12 becomes

$$\mu_X = \mu_F(2w - 1)\frac{1}{k} \tag{7.13}$$

and it is impossible to univocally identify the value of the two parameters w and μ_F because there is an infinite number of pairs of values satisfying this equation. The problem with symmetric force fields persists if one assumes the true bisection model, i.e. $w = 0.5$. In this case, the expected value is

$$\mu_X = \frac{\mu_F 0}{k} = 0 \tag{7.14}$$

and the solution $\mu_X = 0$ is independent from μ_F. However, if $w \neq 0.5$, it is possible to estimate the force threshold value for symmetric force fields by fitting together the systematic and variable errors since w is constrained by the variability of the response (see Eq. 7.11).

For asymmetric force fields, the true bisection model predicts a bias toward the weaker force field as shown by the fact that the bias takes a negative value when k_R is larger than k_L:

$$E[X_{0.5}] = \frac{\mu_F}{k_L k_R}\left(-\frac{1}{2}k_R + \frac{1}{2}k_L\right) = -\frac{\mu_F}{2k_L k_R}(k_R - k_L). \tag{7.15}$$

This equation also shows that the bias depends not only on the stiffness difference $(k_R - k_L)$ between the two sides of the force field but also on the magnitude of the two stiffnesses.

7.4.4 Model Parameters

The free parameters θ of the model consist in the expected value μ_F and standard deviation σ_F of the force threshold, the standard deviation of the positional error σ_P and the weight w of the right limit ($1 - w$ corresponds to the weight of the left limit). The free parameters of the model were fitted separately to the results of both

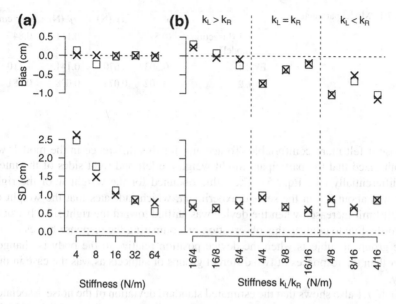

Fig. 7.5 Actual (*empty squares*) and predicted (*crosses*) values for the response bias (*top panels*) and response variability (*bottom panels*). **a** Symmetric force fields (Exp. 1). **b** Asymmetric force fields (Exp. 2)

experiments by minimizing the Mean Square Error (MSE):

$$MSE(\theta) = \sum_{i=1}^{n} (x_i - \mu_X(k_{Li}, k_{Ri}|\theta))^2 + (s_i - \sigma_X(k_{Li}, k_{Ri}|\theta))^2, \qquad (7.16)$$

where x_i and μ_X correspond to the actual and predicted position of the force field centre (Eq. 7.12), s_i and σ_X correspond to the actual and predicted standard deviation, and σ_X the values predicted by the model (Eqs. 7.10 and 7.12), k_{Li} and k_{Ri} correspond to the stiffness in the ith condition of each experiment, and θ the free parameters of the model. The parameters were fitted separately for each experiment. Figure 7.5 shows that the model fits the results of both experiments well.

Table 7.1 reports the parameter values that best fitted the results of both experiments. For Exp. 1, Table 7.1 also reports the estimate for the force and position noises obtained by fitting only the variable error (Eq. 7.11) and by assuming the true bisection model ($w = 0.5$). There were some differences between the values of some parameters which reflected the differences between the results of the two experiments. The most conspicuous difference between the two experiments was the leftward bias of the responses observed in the second experiment. This difference might be associated to a change of position of the device relative to the body midline. In the first experiment, the device was aligned with the body midline while the device was aligned with the shoulder in the second experiment

Table 7.1 Model parameter estimates

	w	μ_F (N)	σ_F (N)	σ_P (cm)
Exp. 1 (bisection model)	(0.5)	–	0.111	0.847
Exp. 1	0.503	0.070	0.141	1.110
Exp. 2	0.302	0.073	0.038	0.781

because it felt more comfortable. To account for this difference in the model, we hypothesized that the participant might weigh the left and right sides of the interval differentially (see Eq. 7.5). The value obtained for the weight w of the right limit was about 0.3 in the second experiment, which indicates that the weight of the left limit increased when the device was shifted toward the right side. It would be worthwhile to confirm this observation by comparing the response biases of the same group of subjects when the device position relative to the body is changed rather than the responses of two different groups of subjects as was the case in this study.

Table 7.1 also shows that the estimated standard deviation of the noise associated with the force signal (σ_F) was about 0.14 N in the first experiment and 0.04 N in the second experiment. In contrast, the estimated standard deviation of the noise in the position signal (σ_P) was in the same range (0.8–1.1 mm) in the two experiments. As it can be seen from Eq. 7.11, the position noise reflects the asymptotic response variability when the stiffness is large while the force noise reflects the increase of the response variability when the stiffness decreases. The parameter values reflect the data: First, the response variability increased less in the second than in the first experiment when the stiffness decreased. Second, the asymptotic values of the response variability were similar in the two experiments when the stiffness increased (see Fig. 7.3).

Finally, fitting the full model to the results of the second experiment yielded an estimate of the minimum force that can be detected by the haptic system. Interestingly, the value of the force threshold was 0.072 N (about 7 g), which is quite close to the results of another study measuring the value of this force threshold directly on the same experimental setup (Baud-Bovy and Gatti 2010a). In this latter study, the haptic device rendered a constant force toward the left or right and the task was to identify the direction of the force transmitted by the handle. The direction of the force was randomly selected while its amplitude varied from 0 to 0.2N (approximately 20 g) The magnitude of the force that was needed to correctly identify its direction in 75 % of the trials ranged from 0.05 to 0.1 N depending on the experimental condition. The order of magnitude of these thresholds is also in line with the results of another study that measured the minimum amount of assistive or resistive force that could be detected during a movement (Zadeh et al. 2008).

7.5 General Discussion

Despite considerable progress in the field of haptic perception [see Lederman and Klatzky (2009), for a recent review], a proper framework to apprehend how sensory signals occurring during the active exploration of an object are integrated and/or combined *within* the haptic modality is still lacking. In particular, as noted in the Introduction, a consensus has yet to emerge about how position and force signals are processed to yield the percept of stiffness.

The main contribution of this chapter is to present a model of how participants might identify the central position of an elastic force field from the information they can sense in the absence of vision. This model makes specific hypotheses about the underlying sensory and integrative processes (see Fig. 7.6a). With respect to the sensory processing stage, this model does not require an estimate of the stiffness to identify the position of the hand when the force threshold is crossed. Instead, the model is based on the idea that the internal state of the sensory system changes when the force threshold is crossed [see Macmillan and Creelman (2005), p. 81–111, for a presentation of Threshold Theory along these lines] and that the hand position is sampled when this event occurs. In Fig. 7.6b, this mechanism is represented by a box that generates a unit impulse when a change in the internal state is detected. The impulse is then multiplied with the hand position signal to yield the position of the hand when the force threshold is crossed. In addition, the sensory stage includes

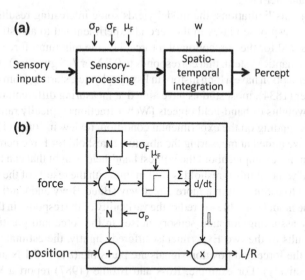

Fig. 7.6 a processing stages of the bisection model. The free parameters of the models are indicated above the blocks. **b** Schematic representation of the sensory processing stage. The two blocks with the letter N represent noise source. The block with the force threshold represents the force detection process that determines the internal state Σ of the sensory system. The last block detects when the internal state has changed. The output of this block indicates the left or right limit of the region of the elastic force field where the force is not perceived

two noise sources (N). The first noise source affects the force signal. It has also an effect on the perceived position of the interval limits because the force threshold will be crossed at different positions depending on fluctuations of the perceived force. The second source of variability affects directly the hand position signal. Finally, the second processing stage corresponds to the spatio-temporal integration of the information acquired during the haptic exploration of the environment. In this model, this stage consists simply in the weighted average of the left and right limits of the interval where the force is not perceived.

The scope of this model is limited. First, this model does not aim at explaining the origin of the force and position signals and how they are processed and integrated at the neuronal level. While it is common to associate position and force signals with the spindles and Golgi tendon organs respectively, it is also known that tactile inputs play an important role in the perception of weak forces [see Srinivisan and LaMotte (1996), Bergmann and Kappers (2008), Baud-Bovy and Gatti (2010b), for some considerations about the possible contribution of tactile and proprioceptive systems in softness, stiffness and weak force perception respectively]. Second, it is not clear how to extend this model to other perceptual tasks, even in the haptic modality. For example, the model does not include signals such as the local deformation of the finger pad that is known to play an important role in haptic shape perception (e.g., Wijntjes et al. 2009). Moreover, the integration stage of this model is extremely simple and does not provide much guidance about how sensory signals might be integrated in another task.

Still, despite its limitations, the model yields some interesting results. First, the analysis of the resp onse biases in the second experiment led to an estimate of the absolute threshold for the perception of a force, i.e., the minimum force that can be detected by the haptic system, that corresponds to 0.7 N or 7–8 gwt (μ_F). While force and weight perception have a long history in Psychophysics, starting from the seminal work of Weber (1834), most studies have aimed at measuring differential thresholds between the weights of hand-held objects [Weber fractions typically range between 5 and 15 % depending on the experimental condition, review in Jones (1986)]. Very few studies have aimed at measuring the absolute threshold for force perception and little is known, for example, about the lightest hand-held weight that can be detected. The estimated value of this threshold is well in line with the results of the few studies that have tried to estimate it in a more direct way (Baud-Bovy and Gatti 2010a, b).

Second, the model was able to predict the variability of the response in the different conditions by assuming constant sensory noises in the force and position signals. While the results of the two Experiments differed slightly, the estimated values of the noise for the force and position signals are plausible ($\sigma_F \approx 0.1$ N and $\sigma_P \approx 8$–10 mm respectively). For example, Ross and Brodie (1987) report a Weber ratio equal to 0.12 for a reference weight of 50 gwt (0.49 N), which corresponds to a correct discrimination threshold of 6 gwt (0.059 N). The threshold was obtained with an adaptive procedure that identified the stimulus intensity that elicited 71 % of correct responses. For a normally-shaped psychometric function, the sensory noise that corresponds to such a threshold is $\sigma_F = DL_{0.71}/\Phi^{-1}(0.71) = 0.059/0.553 \approx 0.1$ N where Φ is the cumulative normal distribution (Gescheider 1997). For smaller weights or forces, the difference threshold and the sensory noise might not decrease much because it is known that the Weber ratio increases with small forces (Jones 1986). Interestingly, the theoretical analysis of this model revealed that the two noise sources are necessary to predict the results of these experiments. As a matter of fact, it was not possible to predict the response variability if one assumed that it derived solely from fluctuations of the force threshold. This analysis tells us something about the information processes operating in this task. In particular, it confirms that participants rely on both position and force signals to identify the central position of an elastic force field.

Finally, this model has some implications for softness perception. As noted in the Introduction, various model of stiffness perception have been proposed. Interestingly, most models don't assume that stiffness is directly perceived. As noted by Srinivasan and LaMotte (1995), tactile cues cannot give direct information about stiffness when exploring the stiffness of a force field by means of a rigid probe because the rate of force change depends on the movement velocity, which must be sensed kinesthetically. In fact, most models of force perception posit that position and force cues must be somehow integrated. In this respect, this model suggests that the perceived stiffness of very compliant objects might be overestimated because the depth of penetration might be underestimated due to an inward shift of the perceived position of the force field boundary. While the implication of this study for stiffness perception is most obvious for models of stiffness perception that rely on the identification of the force field boundary and/or penetration depth in the object, further research is needed to find out the impact of the sensing or perceptual limits of the haptic system on the perception of the stiffness of very compliant objects.

Acknowledgments I would like to thank Francesca Bocca who collected the data in the original study, as well as Netta Gurari, one anonymous reviewer and the editor, Massimiliano Di Luca, for their helpful comments on an earlier version of this chapter.

References

Baud-Bovy G, Gatti E (2010a) Hand-held object force direction identification thresholds at rest and during movement. In: Kappers AML et al (eds) Haptics: generating and perceiving tangible sensations. Proceedings of the international euroHaptics 2010 conference. Lecture notes in computer science, vol 6192, pp 231–236

Baud-Bovy G, Gatti E (2010b) The contribution of proprioception and touch to the perception of weak forces: preliminary results. in: Proceedings of the 26th meeting of the International Society for Psychophysics, Fechner's Day, Padova, Italy, 19–22 Oct, pp 599–603

Baud-Bovy G, Schochia L (2009) Is mass invariant? Effects of movement amplitude and duration. In: Proceedings of 25th Meeting of the International Society for Psychophysics, Fechner's Day, Galway, Ireland, 21–24 Oct, pp 369–374

Beauregard GL, Srinivasan, MA, Durlach NI (1995) The manual resolution of viscosity and mass. In: Proceedings of the ASME dynamic systems and control division, DSC-vol 57–2, IMECE, pp 657–662

Bergmann Tiest WM, Kappers AML (2008) Kinaesthetic and cutaneous cotributions to the perception of compressibility. In: Ferre M (ed) Haptics: perception, devices and scenarios. Proceedings of the international euroHaptics 2008 conference. Lectures notes in computer science, vol 5024, pp 255–264

Bocca F, Baud-Bovy G (2009) A model of perception of the central point of elastic force fields. In: Proceedings of 3rd joint EuroHaptics conference and symposium on haptic interfaces for virtual environment and teleoperator systems, Salt-Lake City, USA, pp 576-581

Brodie E, Ross HE (1984) Jiggling a lifted weight does aid discrimination. Am J Psychol 98(30):469–471

Di Luca M, Knorlein B, Ernst MO, Harders M (2011) Effects of visual-haptic asynchronies and loading-unloading movements on compliance perception. Brain Res Bull 85:245–259

Gescheider GA (1997) Psychophysics: the fundamentals, 3rd edn. Lawrence Erlbaum Associates, Mahwah

Gibson J (1962) Observations on active touch. Psychol Rev 69(6):477–491

Gurari N, Baud-Bovy G (2014) Customization, control, and characterization of a commercial haptic device for high-fidelity rendering of weak forces. J Neurosci Methods (in revision)

Jones LA, Hunter IW (1993) A perceptual analysis of viscosity. Exp Brain Res 94:343–351

Jones LA (1986) Perception of force and weight: theory and research. Psychol Bull 100(1):29–42

Klatzky RL, Lederman S (1995) Identifying objects from a haptic glance. Percep Psychophys 57(8):1111–1123

Lawrence DA (1993) Stability and transparency in bilateral tele-operation. IEEE Trans Robot Autom 9(5):624–637

Lederman SJ, Klatsky RL (1987) Hand movements: a window into haptic object recognition. Cogn Psychol 19:342–368

Lederman SJ, Klatzky RL (2009) Haptic perception: a tutorial. Attention Percep Psychophys 71(7):1439–1459

Macmillan NA, Creelman CD (2005) Detection theory, a user's guide, 2nd edn. Lawrence Erlbaum Associate, Mahwah

Nisky I, Mussa-Ivaldi FA, Karniel A (2008) A regression and boundary-crossing-based model for the perception of delayed stiffness. IEEE Transations on Haptics 1:7382

Parietti F, Baud-Bovy G, Gatti E, Riener R, Guzzella L, Vallery H (2011) Series viscoelastic actuators can match human force perception. IEEE/ASME Trans Mechatron 16(5):853–860

Pressman A, Welty LJ, Karniel A, Mussa-Ivaldi FA (2007) Perception of delayed stiffness. Int J Robot Res 26:1191–1203

Pressman A, Karniel A, Mussa-Ivaldi FA (2011) Perceptual localization and assessment of rigidity. J Neurosci 31(17):6595–6604

Ross HE, Brodie E, Benson AJ (1984) Mass discrimination during prolonged weightlessness. Science 225:219–221

Ross HE, Brodie EE (1987) Weber fractions for weight and mass as a function of stimulus intensity. Q J Exp Psychol 39A:77–88

Salisbury C, Gillespie RB, Tan H, Barbagli F, Salisbury JK (2009) Effects of haptic device attributes on vibration detection thresholds. In: Third joint eurohaptics conference and symposium on haptic interfaces for virtual environment and teleoperator systems (world haptics). pp 115–120

Srinivasan MA, LaMotte RH (1995) Tactual discrimination of softness. J Neurophysiol 73(1):88–101

Srinivisan MA, LaMotte RH (1996) Tactual discrimination of softness: abilities and mechanisms. In: Franzén O, Johansson R, Terenius LY (eds) Somesthesis and the neurobiology of the somatosensory cortex, Birkhauser Verlag, Basel, pp 123–136

Sherrington CS (1900) The muscular sense. In: Schafer EA (ed) Textbook of Physiology vol 2. Pentland, Edinburgh, pp 1002–1025

Thurstone LL (1927) Psychophysical analysis. Am J Psychol 38(3):368–389

Tan HZ, Durlach NI, Beauregard GL, Srinivasan MA (1995) Manual discrimination of compliance using active pinch grasp: the role of force and work cues. Percep Psychophys 57(4):495–510

Weber EH (1834) The sense of touch. Academic Press, London (Original work published 1834)

Wijntjes MWA, Sato A, Hayward V, Kappers AML (2009) Local surface orientation dominates haptic curvature discrimination. IEEE Trans Haptics 2(2):94–102

Zadeh MH, Wang D, Kubica E (2008) Perception-based lossy haptic compression considerations for velocity-based interactions. Multimedia Syst 13:275–282

Chapter 8
Dynamic Combination of Movement and Force for Softness Discrimination

Markus Rank and Sandra Hirche

8.1 Introduction

Softness is an important source of information when interacting with remote or virtual environments (VE) via a haptic human-machine-interface. For example, in telesurgery where the surgeon operates a human-machine interface transmitting his/her actions to a robot performing actions inside the human body, tissue softness can indicate a healthy or non-healthy condition (De Gersem 2005). Humans have no dedicated sense for perceiving softness; instead, inferring an object's compliance haptically requires the combination and integration of information from different sensory sources such as positional cues, force cues, and tactile information—see Chap. 5 for a deeper analysis of mechanisms involved in this process. For many technical systems, including above-mentioned telesurgery setups, tactile cues are not conveyed to the human operator, limiting the information available to infer softness movement and force. In direct interaction with a physical object, the gain and temporal relation of movement and force is determined by the object's mechanical impedance. A telepresence or VE system can alter the impedance by, e.g., time delay in the communication channel (Rank et al. 2010a; Ohnishi and Mochizuki 2007; Pressman et al. 2007; Nisky et al. 2008; Hirche and Buss 2007; Rank et al. 2010; Hirche and Buss 2012) which is found to make participants underestimate stiffness under various circumstances, see also Chaps. 9 and 5. Determining the limits for distortions caused by the technical system that do not affect the operator's percept is crucial to ensure a realistic interaction experience.

M. Rank (✉)
Research Centre for Computational Neuroscience and Cognitive Robotics (CNCR),
University of Birmingham, Edgbaston, UK
e-mail: m.rank@bham.ac.uk

S. Hirche
Institute for Information-Oriented Control, Technische Universität München, Munich, Germany
e-mail: hirche@tum.de

© Springer-Verlag London 2014
M. Di Luca (ed.), *Multisensory Softness*, Springer Series on Touch and Haptic Systems,
DOI 10.1007/978-1-4471-6533-0_8

In the past, perceptual discrimination limits have often been characterized using psychophysical measures such as the just noticeable difference (JND) (Gescheider 1985; Weber 1834), allowing a distinction between perceivable and unperceivable differences in a physical quantity such as a force, length, or impedance by mapping each difference to a proportion in perceptual responses. By simplifying the characterisation of the perceptual system to such a static mapping, valuable information about the time-series characteristics of the environment interaction is lost. Temporal features in the interaction force and movement have though been shown to significantly influence our perception of haptic properties such as hardness (Lawrence et al. 2000) and mass (Baud-Bovy and Scocchia 2009). Perceptual phenomena such as the haptic masking effects found in Rank et al. (2012) could presumably only be understood by looking at the temporal characteristics over time. In softness perception, the amplitude of probing movements was also found to influence human perceptual performance (Tan et al. 1995), a factor that is not accounted for in a softness JND measure. To the authors best knowledge, no conclusive mechanism capturing the combination of movement and force to perceive softness has yet been established.

We propose the usage of dynamic haptic perception models, using differential equations to combine movement and force information together instead of static perception models, e.g., the JND. In this way, the impact of interaction characteristics on the perceptual judgment can be explicitly modelled. Looking at softness perception from a system theoretic point of view, we propose three plausible mechanisms which are capable of discriminating between different soft environments. The detection thresholds predicted by these models vary with the specific interaction movement with the environment. Based on the results from three psychophysical experiments, a dynamic state observer model is identified as a superior prediction model compared to a comparison of identified time delay values and an internal inverse model validation of the body and environment.

Theoretical model candidates from system theory, predicting perception thresholds for temporal misalignment between limb movement and force feedback are introduced in Sect. 8.2. Experimental data from three psychophysical experiments on the perception of time delay in soft, damped and inertial environments are presented in Sect. 8.4, and predictions from the parameterised models are discussed. The chapter is ended with a conclusion on the impact of the results on the design of telepresence and VE systems.

8.2 Perception Model Representations

Perceiving softness generally requires a combination of force and movement cues into a unified percept. Accounting for human perception characteristics in the design, control and evaluation of systems for human-machine interaction such as telepresence or VE systems requires the formulation of quantitative perception models capturing haptic discrimination abilities. The models proposed here are built upon the assumption of an existing *decision criterion* δ. This measure is used to determine which of two response alternatives to choose and can be found in well-established perceptual modelling techniques, e.g. signal detection theory (Macmillan and Creelman 2005) and diffusion models (Ratcliff 1978; Pleskac and Busemeyer 2010).

The perceptual output $\mathbf{y}_p$ at a given response time t_r is determined by

$$y_p(t_r) = \begin{cases} \text{``different''} & \text{if } \exists t \in [0,\, t_r] : |\delta(\cdot)| > \varepsilon, \\ \text{``same''} & \text{otherwise,} \end{cases} \tag{8.1}$$

where ε is referred to as a *decision threshold* and the stimulus onset time is set to $t = 0$.

Remark 1 This formulation of the perceptual process accounts for the fact that in the context of human-robot interaction such as telerobotics, a perceptual decision may be held back and not responded to as soon as the decision has been made. Contrary to Ratcliff (1978), Pleskac and Busemeyer (2010), the formulation of perception models in Eq. (8.1) thus accounts for all decisions made between the stimulus onset up to time t_r.

In most existing computational haptic perception models, $\delta(\cdot)$ is a static function of the sensory input. As an example from softness perception, a static perception model for discriminating two environments with stiffness coefficients k_1 and k_2 could be formulated by setting $\delta(\cdot) = k_1 - k_2$ and setting the threshold value to the JND for stiffness $\varepsilon = \text{JND}_k$. As a consequence, temporal aspects of the interaction such as movement speed, frequency, or interaction duration remain unmodelled. Instead, we use a dynamic modelling approach to capture the decision criterion. We will limit our considerations to ordinary differential equations.

In the following, three perception modelling candidates for the decision criterion $\delta(\cdot)$ in (8.1) are proposed. The main inspiration for these models is drawn from considerations how one would approach the detection of differences in a haptic environment from a system theoretic point of view. Support for the mechanism candidates in terms of neurophysiological and psychophysical evidence is also reported.

8.2.1 Sensorimotor Control Model

The different modelling approaches are discussed using a simplified dynamic model of the human motor apparatus considering only one arm, which is a common simplification throughout the literature (Gil et al. 2004; Yokokohji and Yoshikawa 1994). The state vector $\mathbf{x}_h$ consists of the hand position x_h and velocity $\dot{x}_h$. A block diagram of the arm, controlled to follow a specific state trajectory, is depicted in Fig. 8.1. Note that we make the modelling variables' dependency on time only implicit in favour of a clear presentation. The control mechanism $\Phi_{con}(\mathbf{x}_h, \mathbf{x}_{des})$ determines the forces which must be applied to the limb to follow a desired state trajectory $\mathbf{x}_{des}$. The arm with its mechanical properties $\Psi_{body}(\mathbf{x}_h, \dot{\mathbf{x}}_h, f_{res})$, linearly approximated by a mass-damper system

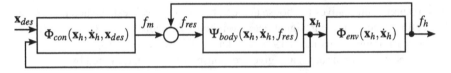

Fig. 8.1 The human arm is abstracted as a state-controlled single joint

$$\ddot{x}_h = -\frac{1}{m_h}(f_{res} - d_h \dot{x}_h)$$

with human-like parameters ($m_h = 2$ kg, $d_h = 2$ Ns/m from Yokokohji and Yoshikawa (1994)) is in contact with the environment. The environment dynamics are contained in $\Phi_{env}(x_h, \dot{x}_h)$ and react to the state x_h with a force f_h. This feedback acts back on the limb and influences the force moving the limb.

Physiologically, humans are equipped with multiple haptic sensors (Hale and Stanney 2004), and we will focus on sensors for the muscle force f_m, limb position x_h and velocity $\dot{x}_h$. Dynamics and noise in the sensory estimates are not considered explicitly, but implicitly respected in the choice of perceptual thresholds $\varepsilon \neq 0$.

8.2.2 Feature Comparison

A straightforward way of discriminating between two soft haptic environments is comparing their characteristic parameters θ. Such parameters include the stiffness coefficient, or, in case a telepresence system including delayed communication is involved, the time delay between movement and force feedback. To be able to compare the two environments on a parameter basis, a system identification technique suitable to capture this specific property must be used, leading to estimates $\hat{\theta}_1$, $\hat{\theta}_2$. Time delay between movement and force could well be identified using an estimate of the covariance between a position input and a force output signal (Ljung 1999). Acknowledging the fundamental assumption of a decision criterion and threshold for perceptual mechanisms in Eq. (8.1), we propose

$$y_p(t_r) = \begin{cases} \text{``different''} & \text{if } |\hat{\theta}_1 - \hat{\theta}_2| > \theta_{thresh} \\ \text{``same''} & \text{otherwise.} \end{cases} \tag{8.2}$$

In studies on monkeys, correlation techniques as a normalised form of covariance methods have been found to be good at explaining brain activity in specific brain regions associated with perception, if the animal attends to a certain visual stimulus (Niebur and Koch 1994). This could be taken as evidence for the existence of a neural substrate for performing correlations efficiently in the brain. Correlation mechanisms can furthermore explain humans' performance in detecting temporal differences in audio-visual signals (Fujisaki and Nishida 2005).

Remark 2 The classical JND measure is defined in the dimension of the physical quantity under consideration, that means the haptic environment property θ (Weber 1834; Jones and Hunter 1990). In that sense, classical perception models are contained in the feature comparison model proposed here and the predictions from the feature comparison model are seen as a baseline for the other dynamic prediction models.

8.2.3 Inverse Model Verification

An alternative approach to judge whether two soft environments have the same or different properties is the use of a model verification technique. In system identification, verification is a standard procedure to check whether an identified system has good generalisation capabilities (Åström and Eykhoff 1971). At first, a haptic environment model is built by exploring one stimulus and identifying its parameters by using, e.g., a covariance method as proposed in Sect. 8.2.2. Secondly, during the exploration of another haptic environment, sensory information is compared to a prediction of the sensory output, given the previously built internal representation of the environment dynamics. If prediction and sensory evidence match, the environments are considered the same. If there is a mismatch between the prediction and feedback, the two environments are classified as different. Diverse verification methods are utilised in various technical applications, differing in the criterion which is taken into consideration for classification.

One possibility for a perception model as proposed in Eq. (8.1) can be formulated based on the force required to move along a specific trajectory. The model

$$y_p(t_r) = \begin{cases} \text{"different"} & \text{if } \exists t \in [0, t_r] : \Delta f_m(t) > \Delta f_{thresh} \\ \text{"same"} & \text{otherwise,} \end{cases} \tag{8.3}$$

is based on the force difference $\Delta f_m(t) = |\hat{f}_m(t) - f_m(t)|$ with $f_m(t)$ being the effective force from all muscles acting on the limb and $\hat{f}_m$ is an estimation of the expected force given the previously identified haptic environment. The decision threshold is denoted Δf_{thresh} in this model. The main difference to the feature comparison model proposed in Sect. 8.2.2 is the fact that the dissociation between a target and a reference environment is not the experimentally varied variable, e.g. stiffness or the communication time delay in a teleoperation system, but the deviant force between the two conditions.

In addition to Eq. (8.3), a perception model based on Weber's Law is proposed, respecting the fact that force discrimination levels have been found to depend linearly on the force level (Tan et al. 1994). A difference between two soft environments can be perceived if the fraction of force error and force magnitude exceeds the Weber fraction w:

$$y_p(t_r) = \begin{cases} \text{"different"} & \text{if } \exists t \in [0, t_r]: \ \Delta f_m(t)/f_m(t) > w \\ \text{"same"} & \text{otherwise} \end{cases} \qquad (8.4)$$

Reconstructing the motor action from a measurement of the state $\mathbf{x}_h(t)$ requires a dynamic model containing the body and the environment impedance. In motor control literature, a model predicting motor actions (force) from an observation of the body state $\mathbf{x}_h$ (movement and position) is referred to as an *inverse model*. There is experimental evidence for the usage of inverse dynamic models in sensorimotor control by predicting the motor actions from the sensed state of the body (Kawato 1999; Shidara et al. 1993). Similarly, an inverse model $\hat{f}_{m,res} = \Phi_{inv}(\mathbf{x}_h)$ capturing dynamics of the arm, sensors and the environment can potentially play a role in perception as well. A stiffness estimation method on the basis of maximum force comparisons between conditions (Tan et al. 1995; Pressman et al. 2007) can be seen as a representative of a perception model using inverse dynamics. Model verifications are closely related to the prediction error method (PEM) which utilises the error between model predictions and sensory information to enhance identification results. This is a well-established technique in system identification (Ljung 1999) and a PEM algorithm has been found to explain the anticipatory perception of sensory events in a plausible way (Szirtes et al. 2005).

8.2.4 State Observer Model Verification

Alternatively to the exerted muscle force $f_m(t)$ as a decision criterion for distinguishing two soft haptic environments, perceptual judgments can be based on the body state $\mathbf{x}_h(t)$. In the proposed model of the arm in Fig. 8.1, consisting of one limb performing a unidirectional movement, $\mathbf{x}_h(t)$ consists of the limb position $x_h(t)$ and velocity $\dot{x}_h(t)$. The resulting haptic perception model is given by

$$y_p(t_r) = \begin{cases} \text{"different"} & \text{if } \exists t \in [0, t_r]: \ |\hat{\mathbf{x}}_h(t) - \mathbf{x}_h(t)| > \Delta \mathbf{x}_{thresh} \\ \text{"same"} & \text{otherwise,} \end{cases} \qquad (8.5)$$

where $\hat{\mathbf{x}}_h(t)$ is a prediction of the body state, given a previously experienced environment dynamics.

A state observer can predict the body state from observations of the motor input and sensory measurements, utilising a forward model of the body and environment dynamics. A state observer with a linear dynamic model is depicted in Fig. 8.2. The estimated dynamics of the limb and environment are contained in the state function $\hat{\Psi}_{body/env}(\hat{\mathbf{x}}_h, \dot{\hat{\mathbf{x}}}_h, f_m)$. Generally, an output function is required to transform states into measurable outputs; however, since humans possess sensors for both position and velocity, no transformation is required here. Comparing the predictions

Fig. 8.2 A block diagram of a state observer

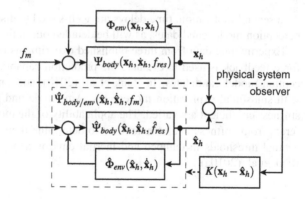

to the actual sensory observations leads to a prediction error which is weighted with a matrix function $K(\mathbf{x}_h - \hat{\mathbf{x}}_h)$ and used to correct future estimates of the body states. In the following, we only consider linear body and environment models and simplify $K(\cdot)$ to a linear matrix multiplication $K(\cdot) = K$. In case $\hat{\Psi}_{body/env}(\hat{\mathbf{x}}_h, f_m)$ captures the body and environment characteristics exactly and the initial state estimate $\hat{\mathbf{x}}_h(0)$ is correct, the state estimate over time $\hat{\mathbf{x}}_h(t)$ equals the real state $\mathbf{x}_h(t)$. If the internal prediction model deviates from the real dynamics because the environment in the second stimulus differs from the comparison condition, the estimated state differs from the real state.

In the case of white noise affecting the output measurement and states, the noise-optimal choice for K is the Kalman Gain. This choice turns the observer into a stationary Kalman filter. Kalman filters have been found to describe sensorimotor control processes well in various situations such as the estimation of hand position (Beers et al. 1999) or posture (Kuo 1995). This is a motivation to consider such a structure as a candidate for perceptual processes as well.

8.3 Model-Guided Experimental Design

A percept of a soft environment can be corrupted in various ways: On the one hand, differences in the stiffness coefficient alter the force feedback magnitude under constant exploration movement; on the other hand, temporal distortions such as time delay between movement and force feedback is capable of completely changing the impression of the environment. Although time delay in haptic feedback is not a natural phenomenon in everyday-life haptic interactions, it is a problem in the operation of telepresence systems over large distances (Peer et al. 2008), e.g., space (Sheridan 1993). We will focus on the investigation of distortions in the haptic combination process due to temporal faults for two reasons: While it is known that time delay between movement and force has a direct impact on the displayed softness (Hirche et al. 2005; Hirche and Buss 2012), the perception of time delay in haptic interaction with an environment is not yet sufficiently understood. However, such knowledge is helpful to provide guidelines and specifications for haptic telepresence systems.

As a second motivation, time delays are well-suited to dissociate between the three perception model candidates, as will be detailed out in the following.

Experimental data from three published experiments on time delay detection in force feedback is used to evaluate the prediction capabilities of the proposed perception model candidates: In Rank et al. (2010), soft environments are explored with sinusoidal exploration movements. Amplitude and frequency as well as the stiffness coefficient are varied. The applicability of the models to environments different from softness is also examined to determine their capability to predict perceptual thresholds in damped and inertial environments as well, using data from Rank et al. (2010a).

8.3.1 Model-Guided Stimulus Selection

The prediction of perceptual thresholds based on the models introduced in Sects. 8.2.2–8.2.4 depends on a multitude of factors, e.g., the interaction movement speed, frequency, and amplitude. Given this high-dimensional parameter space, a fully crossed experimental design with conditions sampled over a range of stimuli is inappropriate. Instead, we choose a model-based selection of experimental stimuli based on predictions for the discrimination threshold of time delay in force feedback from the environment using a linear spring with spring constant k_e. Without loss of generality, the equilibrium point of the spring is set to the position $x_h = 0$. The predicted perception limits of time delay on the basis of the matched filter model and the state observer model depend on the interaction movement $x_h(t)$ with the haptic environment. A sinusoidal movement

$$x_h(t) = A \sin(\omega t) \tag{8.6}$$

with amplitude A and frequency ω is chosen as the interaction pattern since it is easy to understand and perform for participants in a psychophysical experiment. The predictions following from the choice of environment and interaction movement are discussed below.

The force feedback from a soft environment with time delay T_d is expressed as

$$f_h(t) = k_e x_h(t - T_d). \tag{8.7}$$

Respecting the dynamical model of the human arm in contact with the environment illustrated in Fig. 8.1, the overall motor action that is required to move the limb in contact with the environment is

$$f_m(t) = m_h \ddot{x}_h(t) + d_h \dot{x}_h(t) + k_e x_h(t - T_d). \tag{8.8}$$

Without loss of generality, we consider the case that the non-delayed soft environment is explored first. The delayed feedback is perceived second and the sensory evidence from this exploration is compared to predictions from the undelayed stiffness. In addition, we assume that humans have good knowledge of their body dynamics (inertia m_h and damping d_h), and the estimate $\hat{k}_e$ of the environment stiffness coefficient k_e is sufficiently accurate from the non-delayed stimulus exploration.

The inverse model verification model founds on a comparison between sensory observation of the resulting muscular force $f_m(t)$ and the predicted force feedback $\hat{f}_m(t)$. Consequently, $\hat{f}_m(t)$ is determined by

$$\hat{f}_m(t) = m_h \ddot{x}_h(t) + d_h \dot{x}_h(t) + \hat{k}_e x_h(t). \tag{8.9}$$

Setting $\hat{k}_e \approx k_e$ and substituting $x(t)$ with Eq. (8.6), the error between model prediction and sensory feedback is calculated in agreement with (8.3) to

$$\Delta f_m(t) = |k_e A(\sin(\omega t) - \sin(\omega(t - T_d)))|. \tag{8.10}$$

Model verification using a state observer relies on a prediction of the body state

$$\hat{\mathbf{x}}_h(t) = \left[\hat{x}_h(t) \; \dot{\hat{x}}_h(t) \right]^T, \tag{8.11}$$

utilising a forward model of body and non-delayed environment dynamics. The state prediction is the solution of the set of differential equations, expressed in matrix form as

$$\begin{bmatrix} \dot{\hat{x}}_h(t) \\ \ddot{\hat{x}}_h(t) \end{bmatrix} = \begin{bmatrix} 0 & 1 \\ -\frac{d_h}{m_h} & -\frac{k_e}{m_h} \end{bmatrix} \begin{bmatrix} \hat{x}_h(t) \\ \dot{\hat{x}}_h(t) \end{bmatrix} + \begin{bmatrix} 0 \\ \frac{1}{m_h} \end{bmatrix} f_{m,res}(t) + \begin{bmatrix} k_{11} & k_{12} \\ k_{21} & k_{22} \end{bmatrix} \left(\begin{bmatrix} x_h(t) \\ \dot{x}_h(t) \end{bmatrix} - \begin{bmatrix} \hat{x}_h(t) \\ \dot{\hat{x}}_h(t) \end{bmatrix} \right). \tag{8.12}$$

In order to be detectable, the discrepancy in the decision variable must be larger than a threshold variable. In order to determine the amount of time delay between movement and force feedback, the maximum deviance between prediction and sensory observation is to be computed. For the inverse model, the discrepancy is at its maximum at time $\frac{1}{2}T_d$ after the zero-crossings of the predicted (non-delayed) force reference, which is expressed by

$$\Delta f_{m,max} = \Delta f_m(t)|_{t=\frac{1}{2}T_d} = k_e A2 \sin(\frac{1}{2}\omega T_d) \approx k_e A \omega T_d. \tag{8.13}$$

The last step in the calculation holds for small values of ωT_d, which is a valid assumption for the practically relevant range of time delays in telepresence applications and the movement frequencies considered in the experiments.

Similarly, the state observation error can be computed by solving Eq. (8.12) for the specific interaction movement from Eq. (8.6) and the motor action from (8.8). In contrast to the solution for the maximum force error in Eq. (8.13), the maximum state error depends on the entries of the feedback matrix K. These values are unknown.

Fig. 8.3 Six pairs of movement amplitudes and frequencies were chosen in such a way that ω, A and their product $A\omega$ have three different levels respectively

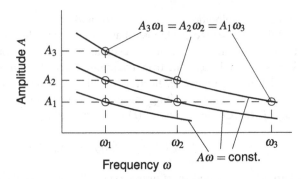

Thus, the experimental conditions are optimized for the inverse model, and the prediction capabilities of the state observer model are tested post-hoc with a feedback matrix K that is identified based on experimental data.

Keeping the time delay T_d at a constant level, the maximum force error as the prediction criterion for time delay detection is higher with a greater amplitude A, and/or higher movement frequency ω. This means in return, that time delay needed to exceed a hypothesized perception threshold on force error is smaller with larger A and/or higher ω. Notably, the maximum force error as introduced in Eq. (8.13) depends on the product of A and ω, predicting that choosing values of A and ω such that their product is constant ($A\omega = $ const.) results in the same detection threshold. For testing the influence of movement amplitude, frequency and their product, a systematic experimental design with three levels for A, three levels for ω and three levels of $A\omega$ as depicted in Fig. 8.3 is chosen.

Another factor in the computation of the maximum force error according to Eq. (8.13) is the stiffness coefficient k_e. The perception model predicts a lower time delay detection threshold in the case where stiffness is higher.

In addition to a soft environment, the prediction capabilities of these models in damping and inertia are explored in order to test a generalisation to other experimental conditions as well. Stimuli with a damping d_e, and an inertia m_e satisfy

$$\left. \frac{\Delta f_{m,max}}{f_m(t)|_{\Delta f_m(t)=\Delta f_{m,max}}} \right|_{d_e} = \left. \frac{\Delta f_{m,max}}{f_m(t)|_{\Delta f_m(t)=\Delta f_{m,max}}} \right|_{m_e},$$

such that the Weber fraction is equal in both conditions, resulting in a constant time delay detection threshold in the case of a perception criterion based on Weber's Law.

8.4 Experimental Investigations

Experimental data from three studies is analysed here. From Rank et al. (2010a), time delay detection thresholds for sinusoidal movements with parameters as depicted in Fig. 8.3 is taken. In addition, detection thresholds for three levels of stiffness under

Table 8.1 Mean detection thresholds (DT) and standard error (SE) of time delay-induced alterations of soft environments depend on the specific interaction movement and the composition of the environment

Condition	Movement variation						Stiffness variation						Environment variation		
	1	2	3	4	5	6	7	8	9	10	11	12	13	14	15
k_e [N/m]	65	65	65	65	65	65	65	65	65	65	65	65	65	0	0
d_e [Ns/m]	0	0	0	0	0	0	0	0	0	0	0	0	0	$\frac{65}{2\pi}$	$\frac{43}{2\pi}$
m_e [kg]	0	0	0	0	0	0	0	0	0	0	0	0	0	0	$\frac{22}{(2\pi)^2}$
$\hat{A}$ [cm]	8.9	10.5	13.2	8.6	10.7	8.74	14.8	14.8	14.4	18.5	18.4	18.7	11.1	11.3	11.2
$\frac{\hat{\omega}}{2\pi}$	0.71	0.70	0.68	0.93	0.92	1.24	1.06	1.08	1.12	0.84	0.85	0.84	1.03	1.05	1.08
DT [ms]	46	47	37	41	37	36	24	25	28	34	31	37	36	15	72
SE [ms]	4.5	7.3	6.3	5.0	4.2	5.8	4.5	6.5	9.3	9.3	4.5	9.5	5.7	2.6	6.1

two different movement patterns are taken from Rank et al. (2010a). Third, the time delay detection thresholds obtained for stiffness are compared to those in damping and inertia environments while keeping the interaction movement constant. This data is reported in Rank et al. (2010). A summary of all experimental conditions and the detection thresholds found in the experiments is provided in Table 8.1. Notably, we also report measurements of participants' mean amplitude $\hat{A}$ and frequency $\hat{\omega}$ of their interaction movement since these have been found to differ from the experimental instructions.

8.4.1 Results

Four substantial findings can be concluded from the experimental findings in Rank et al. (2010a):

1. The detection thresholds for time delay-induced environment alterations are negatively correlated with movement frequency and movement amplitude.
2. Movement amplitude and frequency influence the detection threshold separately.
3. Within the range of experimental conditions, stiffness does not affect perceptual discrimination abilities of time delay in force feedback.
4. A change in the environment due to time delay can be detected easiest in force feedback from a damper, followed by time delay in force feedback from softness. Inertia exhibits the largest detection thresholds.

In order to investigate which perception model candidate is most suited modelling this observed behaviour, parameters for each model are identified and predictions for the detection thresholds are obtained.

8.4.2 Model Predictions

Since experimental methods and the group of participants are not homogenous over the different experiments, we fit mean detection thresholds individually for each experiment. To compare the prediction quality between models, the mean squared error (MSE) is computed. In the following, the individual identification procedures and the prediction results are discussed in detail.

8.4.2.1 Feature Comparison Model

Humans may perceive time delay in a haptic environment per se and compare individual estimates obtained from haptic exploration of the standard and comparison environment. The correlation techniques discussed in Sect. 8.2.2 are indeed well-suited to infer a time delay between movement and force feedback. While an uncertainty in time delay detection performance due to noise in the biological system could lead to a detection threshold different from zero, there is no apparent reason why the uncertainty about the time delay should change with input amplitude, frequency, magnitude, or the type of environment. The predicted time delay detection threshold based on this method is thus constant over conditions. Identification of the only free parameter in this model is achieved by solving

$$\underset{DT^{\theta}}{\arg\min} \ \frac{1}{N_{cond}} \sum_{i=1}^{N_{cond}} (DT_i - DT^{\theta})^2 \qquad (8.14)$$

where N_{cond} is the number of conditions in the respective experiment, and DT^{θ} is the (constant) time delay detection threshold. The solution to this optimisation problem is the mean time delay over all conditions within one experiment. Predictions from this perception model result in a MSE of 127.34 ms^2.

8.4.2.2 Inverse Model Verification

The parameterisation of this model, given the experimental results in Table 8.1 is the result of a nonlinear constrained optimisation problem

$$\underset{DT_i^f, \Delta f_{thresh}}{\arg\min} \ \frac{1}{N_{cond}} \sum_{i=1}^{N_{cond}} (DT_i - DT_i^f)^2 \qquad (8.15)$$

$$s.t. \max \Delta f_{m,i}(t) = \max |f_{m,i}(t) - \hat{f}_{m,i}(t)| = \Delta f_{thresh} \ \forall i \in [1, N_{cond}]$$

where Δf_{thresh} is the (constant) detection threshold for the difference between the delayed and non-delayed exerted force and DT_i^f the corresponding time delay value causing Δf_{thresh}. The predicted motor action on the basis of the measured state $\mathbf{x}_h(t)$ is computed for each individual experimental condition, indexed by i, and denoted $\hat{f}_{m,i}(t)$. A numeric optimisation algorithm based on the interior-point method is used to find the optimal parameterisation fitting all experimental conditions (Byrd et al. 1999). Using the dynamic inverse model to explain average detection thresholds for time delay perception results in lower prediction errors (96.7 ms^2) compared to the feature comparison model prediction. The mean force difference thresholds for the experiments are 1.4 N for the first, 1.2 N for the second, and 1.7 N for the third experiment.

Force difference perception for experiments with slowly-changing forces is known to follow Weber's Law (Tan et al. 1994). The Weber fraction of $\Delta f_h(t)$ could thus be an good model to explain the detection thresholds of time delay as well. The optimisation problem to be solved is similar to Eq. (8.15), namely

$$\underset{DT_i^w, w}{\arg\min} \sum_{i=1}^{N_{cond}} (DT_i - DT_i^w)^2 \tag{8.16}$$

$$s.t. \max \frac{\Delta f_{m,i}(t)}{f_{m,i}(t)} = w \, \forall i \in [1, N_{cond}]$$

with w the Weber fraction. Indeed, the model fit for the experiment with different stiffness levels is admittedly good, with a MSE of only 4.5 ms^2, but the model performs poorly for all other conditions, yielding a total MSE of 127.7 ms^2. Thus, this model performs not better as the feature comparison model being the baseline predictor.

8.4.2.3 State Observer Model Verification

In contrast to the matched filter perception model, the state observer model utilizes an estimation of the body state for the decision about the environment time delay. The difference between the observed state and actual state heavily depends on the choice of the feedback matrix K, as discussed in Sect. 8.2.4. The model predicts perception limits based on a threshold in the state estimation error. The state $\mathbf{x}_h(t)$ consists of two components, namely the limb position $x_h(t)$ and velocity $\dot{x}_h(t)$. While deviations between the observed state and the measured state could be principally based on a generic threshold both on position and velocity, individual models considering a threshold on x_h and $\dot{x}_h$ are considered here:

$$\underset{DT_i^{x_1}, \Delta x_{h,thresh}, K}{\arg\min} \frac{1}{N_{cond}} \sum_{i=1}^{N_{cond}} (DT_i - DT_i^{x_1})^2 \tag{8.17}$$

$$s.t. \max \Delta x_h(t) = \max |x_h(t) - \hat{x}_h(t)| = \Delta x_{h,thresh} \, \forall i \in [1, N_{cond}]$$

and

$$\underset{DT_i^{x2},\Delta\dot{x}_{h,thresh},K}{\arg\min} \frac{1}{N_{cond}} \sum_{i=1}^{N_{cond}} (DT_i - DT_i^{x2})^2 \tag{8.18}$$

$$s.t. \max \Delta\dot{x}_h(t) = \max |\dot{x}_h(t) - \hat{\dot{x}}_h(t)| = \Delta\dot{x}_{h,thresh} \, \forall i \in [1, N_{cond}].$$

The problems formulated in (8.17) and (8.18) have five free parameters to be opti-mised. Due to the comparably low number of experimental conditions which are available for model fitting and the fact that the optimisation problem may indeed be non-convex, the solution can depend on the chosen initial values. Suitable values are found from an initial grid search procedure, meaning a simulation of the state space observer model for different feedback matrices K. Observation errors $\Delta x_h(t)$ and $\Delta\dot{x}_h(t)$ are computed for every candidate of K and the values resulting in the lowest variance for the state error between all conditions of each experiment is taken as initial values for the optimisation problems stated in Eqs. (8.17) and (8.18). Only one feedback matrix K for all experiments is fit to keep the number of variables computationally tractable and reduce the problem of overfitting. However, we do allow for different threshold values $x_{h,thresh}$, $\dot{x}_{h,thresh}$ in the three experiment to account for the differences in experimental methods. As a result, the state observers with feedback matrices

$$K_{x_h} = \begin{bmatrix} 11.8 & 36.3 \\ 33.3 & 31.1 \end{bmatrix}, \text{ and } K_{\dot{x}_h} = \begin{bmatrix} 0 & 9.8 \\ 9.4 & 11.4 \end{bmatrix} \tag{8.19}$$

for predictions based on x_h and $\dot{x}_h$, respectively, give predictions with the lowest mean squared error. Threshold values for the position-based observer are 0.10, 0.02, and 0.07 m. Velocity thresholds are 0.15, 0.04 and 0.07 $\frac{m}{s}$. The MSE values are 98.3 ms^2 for the state observer using the position error as decision variable, and 85.7 ms^2 for the velocity-based threshold. Predictions from all models in all experimental conditions are compared in Fig. 8.4.

8.4.3 Discussion

Comparing the predictions from all models introduced in Sects. 8.2.2–8.2.4 leads to the conclusion that the state observer model with a detection mechanism on the observation error in limb velocity is most successful in capturing the observed percep-tual behaviour. While in the first experiment, conditions with comparable maximum force errors would lead to similar detection thresholds, the inverse model verification method would predict a decreasing detection threshold for an increase in stiffness. However, the second experiment fails to show such behaviour. In general, all dynamic perception models except the model verification model using a threshold based on Weber's Law outperform the static feature comparison model.

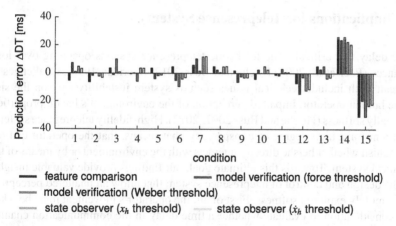

Fig. 8.4 Prediction errors, grouped by experimental condition (1–15, see Table 8.1). Prediction errors are high in environments different than softness (14–15)

The state observer verification model is most successful in predicting detection thresholds for time-delay induced changes in the environmental characteristics, but it also has most degrees of freedom. Claiming the superiority of this model over its alternatives is thus admittedly difficult. Statistical tests such as the Akaike information criterion fail here due to the inhomogeneity of the dataset with respect to participants and methods. However, considering the technical application motivating the perceptual modelling, valuable predictions can still be drawn for the practically relevant set of movement stimuli and haptic environments presented here.

An analysis of the prediction errors in the individual experimental conditions reveals that all proposed models capture the time delay detection thresholds with a significantly lower MSE for the soft environments compared to inertia and damping (Welch's t-test, $t(0.14) = 15.7$, $p < 0.001$). One reason for this lack of generality could be our implicit assumption of an internal representation of the environment that can generate a noise-free and temporally accurate prediction of the reference to the actual sensory feedback. It is known that time perception can be easily disturbed by many factors including attention to the stimulus, the frequency of events occurring etc (Grondin 2010). The difference between the soft, damped and inertial stimuli used in the studies described lies in the relative phase between the position and force signals, thus in their inherent characteristic temporal relation to each other. Modelling temporal uncertainties and noise on the perceptual signals during the exploration may bring further insights into the mechanisms involved in the combination of movement and force into a coherent percept of haptic environments.

So far, all found effects had been attributed to the time delay introduced between position and force feedback. However, using a regular exploration strategy with fixed frequency makes time delay indissociable to a non-linear spring, similar to Leib et al. (2010). The detection could thus as well be a measure of non-linearity in the environment characteristics rather than actual delay. Further studies are required to actually dissociate between these possibilities.

8.5 Implications for Telepresence Systems

Time delay is a critical issue for haptic telepresence systems operating over long distances (Peer et al. 2008; Hirche and Buss 2012; Sheridan 1993). Challenges to be dealt with include technical issues such as system instability and, on the side of the human operator, impaired perception of the environment's haptic properties, especially softness (Hirche and Buss 2007, 2012). High-fidelity telepresence systems must aim for a high degree of transparency, that means, that the operator can not distinguish whether he/she directly interacts with the environment or by means of the technical system. Towards this ultimate goal, our findings provide valuable insights for the design and control of telepresence system that allow an unaltered perception of a remotely explored softness. First of all, the operator's movement must be taken into consideration to evaluate whether a time delay in the communication channel affects softness perception or not. A haptic task which requires only slow movements can tolerate longer delays in the feedback than a highly dynamic task requiring movements with a high frequency. Not only the task can limit the amplitude and movement frequency, but also the haptic interface. A smaller workspace on the one hand, and high friction or uncompensated inertia on the other hand can influence the detection thresholds. The workspace dimensions of the local haptic interface determine the maximum movement amplitude, and detection thresholds increase. With larger inertia and damping of the local haptic interface, the achievable human movement frequency decreases, resulting in a higher detection threshold for time delay.

The finding that a scaling of the stiffness coefficient within the investigated range does not influence the sensitivity of temporal perception is interesting for the application in a specific teleoperation application, namely micromanipulation. In this area, small forces arising in a micro-scale environment must be augmented for the user to provide a perceptible haptic impression (Ando et al. 2001). For the case of delayed haptic feedback, our finding suggests that the scaling factor can be chosen irrespective of haptic latency. Note, however, that we only validated this hypothesis for a limited range of stiffnesses. In extreme scenarios, such as stiff contact with a rigid object, an infinitesimally small time delay may result in an unstable system, which completely changes the characteristics of the system. The human operator may then be able to infer the time delay from increasing oscillations in the force feedback.

Although none of the current model candidates are capable of entirely predicting thresholds for time delay detection in force feedback, the finding of such a dynamic model would have direct application for the design of communication algorithms, or haptic rendering systems as well: The greatest benefit of these models lies in the possibility to consider the influence of interaction movements on the perceptual threshold explicitly. In this way, more accurate predictions whether a time delay in the haptic feedback is perceivable or not can be utilised during the execution of a task, and appropriate measures can be taken, for example in communication Quality-of-Service control algorithms. We take this as a motivation to work further towards this ultimate goal.

8.6 Conclusions and Open Problems

Humans do not possess a dedicated sensor for haptic environment properties such as stiffness, damping, or inertia. Instead, temporal and magnitude information from movement and force feedback must be combined together to infer such measures. System theoretic perception models capable of combining these information sources have been proposed in this chapter. We tested the ability of all model candidates to predict time delay detection thresholds in force feedback. Taking together the results of six psychophysical experiments on time delay perception thresholds, a dynamic state observer model has been identified as the model capturing human discrimination performance best when movement and force feedback are temporally misaligned.

Although all model candidates have been tested for a number of different movements, the pattern was so far restricted to sinusoids of different amplitudes and frequencies. For a more general applicability to haptic telepresence systems, other movements must be considered as well. Ultimately, perceptual responses for time-delayed feedback from arbitrary voluntary explorations shall be predictable. Furthermore, the modelling performance in the third experiment, considering time delay perception levels in stiff, damped and inertial environments have not been captured well by either model proposed so far. Alternative models with other decision criteria could further improve the prediction performance. Together with a dynamic perception model for the influence of magnitude information on the combination of movement and force, conclusions about perception mechanisms for abstract environments containing arbitrary combinations of stiffness, damping and inertia could be eventually drawn.

Acknowledgments This work was supported in part by a grant from the German Research Foundation (DFG) within the Collaborative Research Centre SFB 453 on "High-Fidelity Telepresence and Teleaction". M. Rank is supported by a fellowship within the Postdoc-Programme of the German academic exchange service (DAAD) and the FP7 ICT grant no. 287888 http://www.coglaboration. eu.

References

Ando N, Korondi P, Hashimoto H (2001) Development of micromanipulator and haptic interface for networked micromanipulation. IEEE/ASME Trans Mechatron 6(4):417–427

Åström KJ, Eykhoff P (1971) System identification—a survey. Automatica 7:123–162

Baud-Bovy G, Scocchia L (2009) Is mass invariant? effects of movement amplitude and duration. In: Proceedings of the 25th meeting of the international society for psychophysics, fechner day

Byrd RH, Hribar ME, Nocedal J (1999) An interior point algorithm for large-scale nonlinear programming. SIAM J Optim 9(4):877–900

De Gersem G (2005) Reliable and enhanced stiffness perception in soft-tissue telemanipulation. Int J Rob Res 24(10):805–822

Fujisaki W, Nishida S (2005) Temporal frequency characteristics of synchrony-asynchrony discrimination of audio-visual signals. Exp Brain Res 166(3–4):455–464

Gescheider GA (1985) Psychophysics: method, theory, and application. Lawrence Erlbaum

Gil JJ, Avello A, Rubio A, Florez J (2004) Stability analysis of a 1 DOF haptic interface using the Routh-Hurwitz criterion. IEEE Trans Control Syst Technol 12(4):583–588

Grondin S (2010) Timing and time perception: a review of recent behavioral and neuroscience findings and theoretical directions. Attention Percept Psychophysics 72(3):561–582

Hale KS, Stanney KM (2004) Haptic rendering—beyond visual computing. IEEE Comput Graph Appl 24(2):33–39

Hirche S, Buss M (2007) Human perceived transparency with time delay. Adv Telerobotics 31:191–209

Hirche S, Buss M (2012) Human-oriented control for haptic teleoperation. Proc IEEE 100(3):623–647

Hirche S, Bauer A, Buss M (2005) Transparency of haptic telepresence systems with constant time delay. In: Proceedings of the 2005 IEEE conference on control applications, pp 328–333

Jones LA, Hunter IW (1990) A perceptual analysis of stiffness. Exp Brain Res 79(1):150–156

Kawato M (1999) Internal models for motor control and trajectory planning. Curr Opinion Neurobiol 9(6):718–727

Kuo AD (1995) An optimal control model for analyzing human postural balance. IEEE Trans Bio-Med Eng 42(1):87–101

Lawrence DA, Pao LY, Dougherty AM, Salada MA, Pavlou Y (2000) Rate-hardness: a new performance metric for haptic interfaces. IEEE Trans Robot Autom 16(4):357–371

Leib R, Nisky I, Karniel A (2010) Perception of stiffness during interaction with delay-like nonlinear force field. In: Proceedings of the EuroHaptics conference 2010. Lecture notes in computer science, vol 6191, pp 87–92

Ljung L (1999) System identification, 2nd edn. PTR Prentice Hall, Upper Saddle River

Macmillan NA, Creelman CD (2005) Detection theory: a user's guide. Lawrence Erlbaum

Niebur E, Koch C (1994) A model for the neuronal implementation of selective visual attention based on temporal correlation among neurons. J Comput Neurosci 1(1–2):141–158

Nisky I, Mussa-Ivaldi FA, Karniel A (2008) A regression and boundary-crossing-based model for the perception of delayed stiffness. IEEE Trans Haptics 1(2):73–82

Ohnishi H, Mochizuki K (2007) Effect of delay of feedback force on perception of elastic force: a psychophysical approach. IEICE Trans Commun E90-B(1):12–20

Peer A, Hirche S, Weber C, Krause I, Buss M, Miossec S et al (2008) Intercontinental multimodal tele-cooperation using a humanoid robot. In: Proceedings of the 2008 IEEE/RSJ international conference on intelligent robots and systems, pp 405–411

Pleskac TJ, Busemeyer JR (2010) Two-stage dynamic signal detection: a theory of choice, decision time, and confidence. Psychol Rev 117(3):864–901

Pressman A, Welty LJ, Karniel A, Mussa-Ivaldi FA (2007) Perception of delayed stiffness. Int J Robotics Res 26(11–12):1191–1203

Rank M, Schauß T, Peer A, Hirche S, Klatzky RL (2012) Masking effects for damping JND. In: Proceedings of the EuroHaptics conference 2012. Lecture notes in computer science, pp 145–150

Rank M, Shi Z, Müller S, Hirche H (2010) Perception of delay in haptic telepresence systems. Presence: Teleoperators Virtual Environ 19(5):389–399

Rank M, Shi Z, Müller H, Hirche S (2010) The influence of different haptic environments on time delay discrimination in force feedback. In: Proceedings of the EuroHaptics conference 2010. Lecture notes in computer science, vol 6191, pp 205–212

Ratcliff R (1978) A theory of memory retrieval. Psychol Rev 85:59–108

Sheridan TB (1993) Space teleoperation through time delay: review and prognosis. IEEE Trans Robot Autom 9(5):592–606

Shidara M, Kawano K, Gomi H, Kawato M (1993) Inverse-dynamics model eye movement control by Purkinje cells in the cerebellum. Nature 365(6441):50–52

Szirtes G, Póczos B, Lrincz A (2005) Neural kalman filter. Neurocomputing 65–66:349–355

Tan HZ, Srinivasan MA, Eberman B, Cheng B (1994) Human factors for the design of force-reflecting Haptic interfaces. Dynamic Syst Control 55(1):353–359

Tan HZ, Durlach NI, Beauregard GL, Srinivasan MA (1995) Manual discrimination of compliance using active pinch grasp: the roles of force and work cues. Percept Psychophysics 57(4):495–510
van Beers RJ, Sittig AC, Gon JJ (1999) Integration of proprioceptive and visual position-ianformation: an experimentally supported model. J Neurophysiol 81(3):1355–1364
Weber EH (1834) Die Lehre vom Tastsinne und Gemeingefühle auf Versuche Gegründet. Friedrich Vieweg und Sohn, Braunschweig
Yokokohji Y, Yoshikawa T (1994) Bilateral control of master-slave manipulators for ideal kinesthetic coupling—formulation and experiment. IEEE Trans Robotics Autom 10(5):605–620

Chapter 9
Perception of Stiffness with Force Feedback Delay

Ilana Nisky, Raz Leib, Amit Milstein and Amir Karniel

9.1 Introduction

Throughout the course of our everyday lives, we retrieve information from our environment and generate internal representations of the world around us (Karniel 2009, 2011; Kawato 1999; Wolpert and Kawato 1998; Wolpert et al. 1998). The sense of touch helps us to generate internal representations of the mechanical properties of objects, and we use it both for constructing perception and for guiding action. For example, a surgeon may palpate a tissue and use the perceived stiffness for diagnosis, but also for determining how strongly to grip a scalpel while cutting it. This book is focused on understanding how the human sensorimotor system integrates various sources of information to form a representation of stiffness—the linear relation between position and force. In this chapter, we will examine attempts to answer this question when users interact with artificially changed environment in which the force resulting from an interaction with the object is delayed, such as in the case of remote bilateral teleoperation.

Bilateral teleoperation allows human operators to interact with distant objects by moving a local robotic device and sensing the forces reflected from a remote device that interacts with the environment. Successful bilateral teleoperation can improve various aspects of our lives, including the practice of telemedicine, such as telesurgery (Anvari 2007; Marescaux et al. 2001; Satava 2006) and remote rehabilitation (Duong et al. 2010; Reinkensmeyer et al. 2002), the safe handling of hazardous materials, the ability to perform space vessel maintenance tasks (Hirzinger et al. 1993; Imaida et al. 2004; Reintsema et al. 2007; Yoon et al. 2004), and the addition of a personal touch to standard telecommunications (Avraham et al. 2012; Karniel et al. 2010). It

I. Nisky (✉)
Department of Mechanical Engineering, Stanford University, Stanford, CA, USA
e-mail: nisky@bgu.ac.il

R. Leib · A. Milstein · A. Karniel
Department of Biomedical Engineering, Ben-Gurion University of the Negev, Beer-Sheva, Israel

© Springer-Verlag London 2014   167
M. Di Luca (ed.), *Multisensory Softness*, Springer Series on Touch and Haptic Systems,
DOI 10.1007/978-1-4471-6533-0_9

can also facilitate telementoring in fields where motor skills acquisition might benefit from a telepresent teacher (Gillespie et al. 1997; Sheridan 1997), such as in medicine (Ballantyne 2002; Rosser et al. 1997), sports, performing arts, and education.

Transparency quantifies the fidelity of a teleoperation system, and is typically defined as the ability to accurately display remote environment properties to the operator, and the ability to accurately execute the movements of the operator in the remote environment. Most of the classical transparency studies have focused on the optimization of teleoperation system in isolation from the human operator and the environment—a system-centered approach. Recently, an alternative, human-centered, approach to this problem, that emphasizes the subjective experience of the operator and the successful outcomes of the operation was suggested (Nisky 2011; Nisky et al. 2008a, 2011, 2013). A key step in developing a human-centered approach to remote teleoperation is to explore the effects of delay on the perception and action of the operator, and in this chapter, we will address the implications of delay effects on the design and analysis of teleoperation systems.

Perception, action, and adaptation to externally induced delay are also interesting to explore because they can reveal how the human motor control system is able to successfully cope with internal delays that change during the course of our life. In addition, understanding mechanical interaction with delayed environments may advance the knowledge of how sensory force feedback is integrated with feedback about executed movements for control and for estimation of mechanical properties, even when force is not delayed. During interaction with spring-like elastic objects, there is a linear coupling between movement and force, and therefore, it is impossible to disambiguate which information is used by the motor system in various contexts. Delay of force feedback breaks this coupling between motion and force, and generates a haptic illusion. Various visual and haptic illusions (Bicchi et al. 2008; Gentaz and Hatwell 2004; Hogan et al. 1990; Lederman and Jones 2011) are often used in neuroscience as tools to explore underlying neural mechanisms by suggesting computational models that can explain the behavior of participants under these illusions. We will discuss the implications of the perceptual effects of delay on the understanding of sensorimotor processes in the brain, and in particular, on the combination of force and position control in palpation of soft objects.

Perception of stiffness can be explored using real elastic objects or using virtual elastic springs rendered by means of haptic devices (Biggs and Srinivasan 2002). The advantage of using virtual environments is that when a programmable robotic device generates the mechanical coupling between motion and force, the researcher is free to design the environment such that each component of the mechanical interaction can be studied separately. Users may face various environments, such as elastic force fields with delay (Nisky et al. 2008b, 2010; Pressman et al. 2007, 2008), viscous force fields (Shadmehr and Mussa-Ivaldi 1994), or various nonlinear force-position (Leib et al. 2010; Mugge et al. 2009; Nisky et al. 2010, 2011; Repperger et al. 1995) and force-velocity (Millman and Colgate 1995) relations. Utilizing interaction with virtual environments is critical for exploring the effect of delay on perception and action, because it allows easy implementation of a pure delayed spring without the need of setting up a teleoperation channel, and hence avoiding jitter, communication distortion, and controllers' dynamics.

This advantage of using virtual environments comes with the price of potential inaccuracies in the rendered environments resulting from hardware and control limitations. These limitations, however, may be mitigated by experimental protocol design. For example, the effect of delay is typically evaluated by comparison between delayed and non-delayed elastic objects in a forced-choice paradigm rather then using magnitude estimation psychophysical methods.

To study the effect of delay on perception and action, participants are often asked to interact with a force field that has the mechanical properties of a one-sided spring, i.e., the applied force is proportional to the penetration into the field:

$$f(t) = \begin{cases} k\,(x(t) - x_0)\,, & x(t) \geq 0 \\ 0, & x(t) < 0 \end{cases} \tag{9.1}$$

where k is the stiffness of the elastic force field, and x_0 is its *nominal boundary*. A one-sided spring is a first order approximation of the mechanical properties of many real-life objects that apply forces on the user only when compressed. When a delay of Δt ms is introduced, the force is proportional to the penetration Δt ms ago, as in:

$$f(t) = \begin{cases} k\,(x(t - \Delta t) - x_0)\,, & x(t - \Delta t) \geq 0 \\ 0, & x(t - \Delta t) < 0 \end{cases}. \tag{9.2}$$

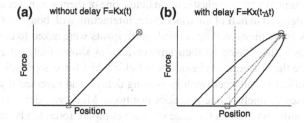

(a) without delay F=Kx(t)   **(b)** with delay F=Kx(t-ₐt)

Fig. 9.1 Force-position plane trajectories of one forward and back palpation cycle in contact with a spring-like field without (**a**) and with (**b**) delay. A sinusoidal movement trajectory (as a function of time) was assumed here for illustration purpose. The *solid vertical line* denotes the position of the *nominal boundary*. The *dashed vertical line* indicates the position of the *effective boundary*. The *square symbol* represents a possible *perceived boundary*. The 'o' and 'x' symbols represent the maximum force and penetration, respectively. In the case of a delayed spring-like field (**b**), the operator first penetrates the field without force feedback, and only after the delay does the force increase gradually. As the operator reverses the movement direction (at the 'x' mark), the force continues to increase for the duration of the delay and then decreases (at the 'o' mark). The slope of the *dotted-dashed line* represents the nominal, non-delayed, stiffness, and the *dotted line* represents the perceived stiffness based on the model in Pressman et al. (2007) that estimates perceived stiffness as the ratio between peak force and perceived penetration, i.e. the distance between the position at the time when the peak force was applied, and the perceived boundary

Stiffness is defined as the ratio between applied force and penetration. In the non-delayed case, the force-position trajectory is a straight line (Fig. 9.1a); however, delay causes this trajectory to become elliptical (Fig. 9.1b), and the force is no longer a single-valued function of the position. During the probing movement, the local stiffness is at some times lower and at other times higher than the non-delayed stiffness. In addition to the *nominal boundary*, an *effective boundary* can be defined, which is the position where the forces are first applied on the user. The effective boundary of a linear spring-like field is a region where stiffness is ill defined, i.e., the derivative of the force-position relation exhibits a sudden transition from zero to a non-zero value, and at the transition point, it has different values along different directions. In the non-delayed case, the effective boundary is the nominal boundary. In the delayed case, they do not coincide, and the exact position of the effective boundary depends on the delay and on the velocity of the probing movement. In general, the faster the users move their hand while interacting with the elastic force field, the larger the effects of delay on force-position trajectories are. It is also important to keep in mind that larger movement velocities also enhance device-related artifacts such as the effects of device inertia and sampling time interval.

9.2 The Effect of Delay on Stiffness Perception

To systematically explore the perceptual effect of delay, Pressman et al. (2007) used a forced choice paradigm with the method of constant stimuli. In each trial, the participants were presented with two elastic force fields: one of them was never delayed and the other was delayed in half of the trials. After interacting with both elastic fields for as long and as many times as they wanted, participants were asked to answer which of the two was stiffer. Based on their answers, it was shown that participants tend to overestimate the stiffness of elastic force fields when force feedback is delayed, and that this effect is enhanced with increasing delays, and reversed if direction of delay is reversed (namely, force precedes position). Moreover, delay caused over-estimation of stiffness even in the case of more complex force fields, in which the delay was present either only during the inward portion of the probing movement (and the forces applied during the outward portion of the probing simulated a linear, non-delayed, elastic force field), or only during the outward portion (and accordingly, the forces applied during the inward motion simulated a non-delayed elastic field). In this experiment, participants made planar movements of their entire arm including shoulder, elbow, and wrist joints in the sagittal plane passing through their shoulder (Fig. 9.2a) while interacting with a planar robotic manipulandum (Shadmehr and Mussa-Ivaldi 1994), and the interaction forces were applied in the sagittal plane toward them. The participants were free to cross the boundary of the elastic field as much as they liked.

To explain these experimental results, the authors suggested that the perception of stiffness is formed based on a limited subset of force-position information that the participant experienced during probing. Their best model for this effect suggested that

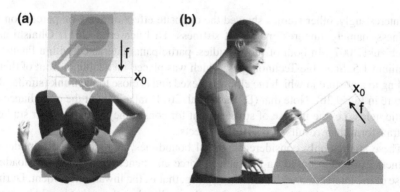

Fig. 9.2 In studies of perception of stiffness with delay, seated participants interact with robotic devices that emulate interaction with one-sided elastic force field. In both panels, the *orange gradient* represents the strength of the elastic force field. The *arrow* denotes the direction of the applied force, f, and x_0 is the location of the *nominal boundary* of the elastic force field. In **a**, a view from above of the posture in Pressman et al. (2007, 2008, 2011) is depicted. In **b**, a side view of the posture in Nisky et al. (2008b) and in the "elbow" condition in Nisky et al. (2010) is depicted

the participants' answers could be explained by the ratio of the maximum interaction force and the *perceived penetration* into the force field. The perceived penetration is calculated relative to the *perceived boundary* of the elastic field, which is estimated by a combination of the *effective boundary* at a specific trial with a prior based on the effective boundary positions that were experienced in previous interactions with the elastic object. During the experiment, participants interacted with delayed and non-delayed elastic force fields, and hence, experienced both shifted and identical to nominal effective boundaries. Hence, the perceived boundary is between the nominal and effective boundaries (as depicted in Fig. 9.1b), and the perceived penetration is larger than the effective one, but it is smaller than the nominal. The maximal force is a function of the nominal penetration, and hence the stiffness is overestimated, as depicted in Fig. 9.1b.

To further test the effect of the perceived boundary location on stiffness perception in cases other than delay of force feedback, Pressman et al. (2008, 2011) performed two additional studies. In these studies, they caused a shift in the position of the perceived boundary by occasionally shifting the entire elastic object away from or towards the participant. When the boundary of an elastic force field was shifted away from the participants they underestimated its stiffness, whereas a shift toward the participant caused overestimation. In this case, unlike in the case of delayed force field, the *effective* and *nominal* boundaries coincide, and both are shifted. The prior estimation of the position of the boundary is based on the entire history of interaction with the shifted as well as regular elastic force fields, and therefore, the *perceived* boundary is shifted less than the effective and nominal, As a result, the perceived penetration is larger in the case of an away shift, and smaller in the case of a toward shift, causing underestimation and overestimation of stiffness, respectively.

Interestingly, other studies showed the opposite effect of delay on perception of stiffness, namely, underestimation of stiffness (Di Luca et al. 2011; Ohnishi and Mochizuki 2007). In both of these studies, participants interacted with a Phantom Premium 1.5, Sensable Technologies, which was placed on a table in front of them, leading to a posture in which the elbow is flexed and is close to the trunk (similar the posture in Fig. 9.2b). Note that (Di Luca et al. 2011) talked about the compliance of elastic fields, i.e. the inverse of stiffness, but for consistency with the other studies, we transform their results into stiffness effects.

These studies only considered nominal boundaries, and suggested a model of stiffness estimation based on the ratio of force and penetration during the loading phase of the interaction with the elastic spring, that is, the inward movement. During the loading phase, the forces are consistently smaller than those that would be experienced for similar penetration without delay, whereas during the unloading phase, the forces are larger. Therefore, a higher weight on the information that is sensed during the loading phase can explain underestimation of stiffness due to delay. This model assumes that the higher weight of the loading phase is a result of a statistically optimal integration of information between the two phases, and is supported by the result that the just noticeable difference for stiffness discrimination is larger when users interact with the spring in the unloading phase only when compared to loading phase only (Di Luca et al. 2011). In addition, when the visual information of movement was delayed, the effect was reversed. This is consistent with the idea that the users combined visual and proprioceptive information to obtain an estimate of the penetration into the elastic field (Ernst and Banks 2002), and therefore, delaying the visual information effectively reverses the sign of delay when compared to delay in force.

While it is clear from the studies that we have reviewed so far that temporal mismatch between force and position information during interaction with elastic force fields biases the perception of stiffness, the results are unequivocal about the direction of this bias. In addition, in Pressman et al. (2007), the effect of delay was similar regardless to whether it was introduced during the loading or unloading phases of exploration, which is not consistent with the models that were suggested in Di Luca et al. (2011) and Ohnishi and Mochizuki (2007). There are, however, numerous differences between the ways the participants interacted with the elastic force fields in these studies, including: properties of the haptic devices that were used for rendering the elastic force fields (the relatively powerful and large robotic manipulandum compared to the smaller Phantom Premium 1.5 haptic device); the posture of participants; the joints and muscles that were active during interaction with the elastic force field; the direction of the applied forces with respect to gravity, and the trajectories of their exploratory palpations (Kaim and Drewing 2008). Moreover, in Di Luca et al. (2011), users received both haptic and visual information during interaction with the virtual objects, whereas in Pressman et al. (2007), they received only partial visual information about their lateral movement which was orthogonal to the direction of applied forces. These factors are very likely to be at least partially responsible for the discrepancy in the results.

Nisky et al., studied how two of these factors may influence the way delay biases perceived stiffness:

- The effect of the trajectories of exploratory palpations, focusing on crossing the boundary of the elastic force field (Nisky et al. 2008b).
- The effect of the joint that pivots the palpation movements (Nisky et al. 2010).

They found that the perceptual effect of delay depends on the probing: participants underestimated delayed stiffness when they did not move across the boundary of the elastic field and maintained continuous contact with the field, and overestimated delayed stiffness when they frequently crossed the boundary and left the field (Nisky et al. 2008b). To explain these effects, the authors suggested a model that, unlike the previously described models, takes the entire information acquired during interaction with the elastic field into account. According to this model, perception of stiffness is constructed from a convex combination of the inverse of the slope of regression of position over force (which predicts overestimation of stiffness), and the slope of regression of force over position (which predicts underestimation of stiffness). The fraction of the probing movements that included boundary crossing out of the overall probing movements within each trial determined the weight of position over force regression.

The authors suggested that this model might imply that the estimation process is related to the control policy that guides the movement of the hand (as illustrated conceptually in Fig. 9.3a). When the hand moves in a homogeneous environment, such as in the case of probing an elastic force field continuously, position control

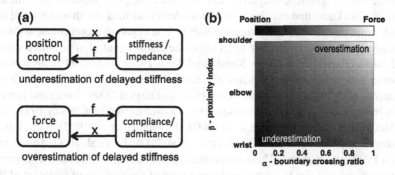

Fig. 9.3 The force and position control hypothesis. **a** A conceptual model of the control and estimation causalities. The physical environment may be thought of as impedance (stiffness) or admittance (compliance) and the appropriate estimation is determined by the control of probing movements. **b** Both control causalities are combined in probing of elastic force fields, and they are weighted according to the frequency of crossing the boundary of the elastic force field, and according to the proximity of the joint that is used for probing. This is reflected in the perceptual biases due to delay. The *shade* represents the weighting of the control causality (*dark*—position control, *light*—force control) as well as the perceptual effect of delay (*dark*—underestimation, *light*—overestimation)

dominates the control of probing interaction. In this case, sensed force is a linear function of commanded displacement; namely, the environment is impedance, and the appropriate estimation of stiffness is the slope of force over position regression. However, when the hand encounters the boundary frequently and the participant experiences abrupt changes in stiffness, the preferred control causality becomes force control. In this case, sensed displacement is a linear function of the applied force; namely, the environment is admittance, and stiffness is estimated based on the inverse of the slope of position over force regression. This idea was further supported in a study where perception of the stiffness of a stepwise linear force-position relation was successfully explained as the inverse of estimation of compliance (Leib et al. 2010). In this study, the stiffness of the force field increased several times during a single probing cycle, and its results suggest that probing of an object with multiple boundaries in which the levels of stiffness increase is dominated by force control.

The distinction between force and position control is prominent in robotics (Raibert and Craig 1981). Recent evidence suggests that this may also be relevant for the human motor system. While there are some studies that report independent motion and force control (Chib et al. 2009; Venkadesan and Valero-Cuevas 2008), it is more likely that in the motor system they are weighted gradually rather than switched in a discrete manner. This weighting may depend on the impedance of the environment, as supported by the findings that increasing environment stiffness elevates the weight of force contribution to sensory estimation (Mugge et al. 2009) and causes transition from restoring an unperturbed trajectory to compliance with the perceived object boundary (Chib et al. 2006).

In Nisky et al. (2010), the authors hypothesized that the effect of delay on perception of stiffness might depend upon the joint that is used to pivot palpation movements. Intuitively, we know that our fingers are more dexterous than our shoulders, and our shoulder muscles stronger than finger muscles. The apparent biomechanical differences between limb segments are reflected in the distinct control of proximal versus distal joints (Kandel et al. 2000; Kurata and Tanji 1986), as evident in anatomical (Brouwer and Ashby 1990; Davidson and Buford 2006; Lemon and Griffiths 2005; McKiernan et al. 1998; Palmer and Ashby 1992; Riddle et al. 2009; Turton and Lemon 1999), functional (Biggs and Srinivasan 2002; Domenico and McCloskey 1987; Hall and McCloskey 1983; Hamilton et al. 2004; Refshauge et al. 1995), and clinical (Colebatch and Gandevia 1989; Dijkerman et al. 2008; Lu et al. 2000; Turton and Lemon 1999) observations. These differences might be responsible for the superiority of the distal muscles in control and perception of the position of endpoint of the limb (Domenico and McCloskey 1987; Hall and McCloskey 1983; Refshauge et al. 1995; Tan et al. 1994) and the superiority of proximal muscles in the control of force (Biggs and Srinivasan 2002; Hamilton et al. 2004).

Indeed, in an experiment in which the users did not have access to the boundary of the elastic force field, they found a proximodistal gradient in the amount of underestimation of delayed stiffness in the transition between probing with shoulder, elbow, and wrist joints. That is, the underestimation was largest when users used their wrist, and was smallest when they used the shoulder. In some of the participants, the effect when probing with the shoulder was even overestimation. To explain these results, the authors used the same regression-based model, but in this case, the weight of the estimation of compliance (that is related to the force controller) increased with the proximity of the probing joint; that is, when probing was pivoted at the shoulder, participants weighted the compliance estimation more, and when probing was pivoted at the wrist, they relied on stiffness estimation.

To summarize, the regression-based model implies that if perception is directly related to the control policy that guides the hand, estimation of stiffness is related to position control, whereas the estimation of the inverse of compliance is related to force control. If this is true, then the findings about the effect of delay on perception of stiffness provide an indirect evidence for the combination of position and force control in explorative movements. The findings in Leib et al. (2010), Nisky et al. (2008b), Nisky et al. (2010) suggest that position and force control are weighted commensurately with the demands opposed by the environment, such as boundary crossing and with the demands opposed by the biomechanical and control constraints of the motor system itself, such as the proximity of the joint that is involved in the probing movement. The weight of force control is increased as the boundary crossing ratio increases and with the proximity of the probing joint. Therefore, a proximodistal gradient in underestimation of delayed stiffness, as well as a transition between underestimation and overestimation of delayed stiffness when increasing boundary-crossing ratio are observed (Fig. 9.3b).

Now, let us return back to the inconsistency in the reports about the direction of the bias in perceived stiffness due to delay, while keeping in mind that constraining the probing movement to the shoulder joint caused the effect of delay to be shifted towards the direction of overestimation. Based on these results, we suggest that in the study of Pressman et al. (2007), the involvement of the shoulder, together with the repeated crossings of the boundary of the elastic field during palpation [see Fig. 9.2 dashed traces in Pressman et al. (2007)], drove the users to weight force control higher than position control, and was responsible for stiffness overestimation due to delay. Another factor that could have contributed to increasing the weight of force control, and consequently, to overestimation of stiffness due to delay is the relatively high stiffness levels of the explored objects. In Pressman et al. (2007), users interacted with stiffness levels ranging from 150 to 600 N/m, compared to levels up to 50 N/m in Di Luca et al. (2011), Ohnishi and Mochizuki (2007). Indeed, in Nisky et al. (2008b), where the stiffness levels were intermediate, ranging between 85 and 265 N/m, mixed results of over- and underestimation of delayed stiffness were observed. Such reasoning is consistent with other findings, where during interaction with elastic force fields, with increasing stiffness level, users tend to weight force cues stronger in control (Chib et al. 2006) and in sensory integration (Mugge et al. 2009).

To yield perceptual effects, the delay does not need to be large. In all the studies reviewed here, the delay between force and position ranged between 5 and 60 ms. In a different study (Ishihara and Negishi 2008), delay was reported to have perceptual effects at values as small as 4 ms. Larger delays are typically not studied because they disrupt palpation movements, and because the perception of an elastic spring as a physical object breaks down when the delay becomes too large relative to the typical palpation frequency of around 2 Hz (Brown et al. 2004; De Gersem et al.

2005; Karniel et al. 2010). Importantly, such delays have a practical meaning—they are typical for long distance teleoperation on earth [e.g. "Operation Linbergh" in which surgery was performed via teleoperation between New-York and Strasbourg (Marescaux et al. 2001)] or for ground-to-earth-orbit teleoperation via radio link (Reintsema et al. 2007).

Finally, while we focused on perception of stiffness, other mechanical properties of objects play an important role in our representation and interaction with the world. Delay was shown to induce changes in the perception of many other mechanical properties, such as mechanical impedance, including mass, viscosity, and stiffness (Hirche et al. 2005; Hirche and Buss 2007), and texture, including roughness and friction (Okamoto et al. 2009).

9.3 The Effect of Delay on Action: Motormetric Representation of Stiffness

So far, we have discussed the effect of delay on perception of stiffness as can be measured by asking participants about the relative stiffness of different objects. Such perceptual aspect is useful for verbal communications and for decision-making. For example, it can be used in the daily task of choosing fruits at the grocery store, or in medical applications, where the stiffness of the tissue conveys to the physician important information about its health. Internal representation of the mechanical properties of an object, however, plays at least an equally important role in action, or, more specifically, the planning and execution of movement in contact with the object. Examples are when a surgeon determines the necessary forces to apply on a surgical scalpel for cutting a tissue, or when an artist uses their fingers to remove just enough material while shaping a sculpture.

Arguably, the most adequate way of describing the relation between perception and action is a bidirectional coupling. Perception of the mechanical properties of the environment is important in planning future actions, and at the same time, natural haptic exploration of the environment is active—we move and probe the environment to create haptic perception. Inconsistencies between perception and action are evident in many tasks. For example, in adaptation to force fields that are applied by a robotic device (Shadmehr and Mussa-Ivaldi 1994), participants report that by the end of training, they can no longer feel the field, and when the force field is suddenly removed, they report that they begin to feel a sensation of an opposite force even though the robotic device is not applying any forces. Several studies have reported resistance of motor actions to visual illusions (Aglioti et al. 1995; Carey 2001; Ganel and Goodale 2003; Goodale and Milner 1992) and inconsistency between perception and action in various tasks (Aglioti et al. 1995; Ganel and Goodale 2003; Goodale and Humphrey 1998; Goodale and Milner 1992). Such dissociation between the motor and perceptual responses was also reported with regard to the size-weight illusion (Brayanov and Smith 2010; Flanagan and Beltzner 2000).

To quantitatively assess the effect of delay on action-related representation, a modification to the perceptual forced choice paradigm was developed that is based on adaptation to force fields and aftereffects in catch trials (Pressman et al. 2008). Instead of processing the answers of participants about their subjective perception using a psychometric analysis (Wichmann and Hill 2001), the authors have introduced the term *motormetric* analysis to designate procedures that are based on observable motor actions for assessing the evolution of perceptual models (Pressman et al. 2008).

In the psychometric analysis, in each trial, participants compared pairs of force fields, and their judgment about the relative stiffness was used to construct a psychometric curve. In the motormetric analysis, a block of trials was used to extract the motor response to a difference between a pair of force fields. In each trial, participants were asked to reach out and back to a target inside a virtual force field. In each block, the participants first performed several trials and adapted to reach into a certain force field that could be delayed or not. Then a catch trial was presented, in which the force field was unexpectedly replaced with a non-delayed field with a different level of stiffness (as depicted in Fig. 9.4 until the 'catch' trial).

The amount of overshoot or undershoot is a proxy to action-related overestimation or underestimation, respectively, of the trained force field relative to the catch field, and it was used to construct the motormetric curve as a function of the difference in stiffness levels between the trained and the catch force fields.

Using this paradigm, the authors showed that perceptual and motor-related estimations of stiffness are inconsistent. While the direction of the perceptual bias was toward overestimation of stiffness, users did not reach the target, i.e. underestimated the stiffness of the elastic force field. Moreover, in the same study, similar motor responses were observed in an experiment where instead of introducing delay, they shifted the elastic force field away from the user. Interestingly, the perceptual effect of the shift was in the opposite direction from the effect of delay—participants underestimated the stiffness when the force field was shifted away from them.

This paradigm was further refined to allow exploring the effect of delay on perception and action in the same experiment (Nisky et al. 2011). To do this, the authors incorporated questions about the relative stiffness of the last two force fields the

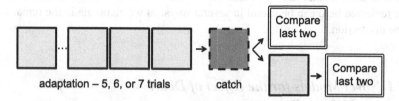

Fig. 9.4 The structure of a single block of the combined motormetric and psychometric experimental protocol. Each block included five to seven training trials with either a delayed or a non-delayed elastic force field (*box* with *thick line*), followed by one of ten possible non-delayed catch trials (*box* with *dashed line*), followed by a question: "Which of the last two fields had higher level of stiffness?" (*box* with *double line*), or a repetition of the trained field and a question

user has interacted with during the experiment (Fig. 9.4). Using this paradigm, they found an inconsistency between the effect of delay on perception and action in virtual teleoperated needle insertion—delay caused motormetric underestimation without changing the perception of stiffness. For the purpose of our discussion here, the virtual teleoperated needle insertion is just an example for a nonlinear elastic force field. Therefore, we do not elaborate here about the needle insertion-related aspects of this study, and refer the interested reader to (Nisky et al. 2011).

This evidence might indicate two parallel neural processes, but the studies we reviewed here did not provide direct support of this hypothesis. Therefore, further exploration of this intriguing gap is open for future studies. Interestingly, while not providing an explanation for the underlying mechanisms that are causing this inconsistency, its experimental evaluation was successfully used for improving transparency of a simple virtual teleoperation channel (Nisky et al. 2011), as will be discussed in more details below. More generally, the effect of delay on motor performance was mostly studied in the context of teleoperation, such as a virtual bimanual pick and place task (Cooper et al. 2012), or a peg in a hole task (Yip et al. 2011). In these studies, however, the focus is on task performance rather than on trying to pinpoint the specific contribution of the mechanical properties of objects, and in most cases, the objects are rigid.

9.4 Discussion

We described three conceptually different explanations for the complex effects of delay on perception of stiffness. This means that all of them are very likely to be wrong. This, however, does not discourage us, because in fact, "all models are wrong, but some are useful" (Box and Draper 1987). A false but simple model can be very useful if it provides an understanding for what exactly about the model is wrong (Fernandes and Kording 2010), if it generates working hypotheses for future experiments that may refine or refute the model and suggest alternative explanations for the behavioral results, or if it provides a sufficiently good approximation under certain conditions that have practical application. We think that the models that we have reviewed here can be useful in several ways, as we elaborate in the remainder of the discussion.

9.4.1 Novel Models for the Effect of Delay on Stiffness Perception

New models may be generated from the combination of two or more of the models that we have described here. The regression model (Nisky et al. 2008b, 2010) could be combined with the perceived boundary model (Pressman et al. 2007, 2011) into a single model that calculates a regression line that is constrained to pass through the perceived boundary. This model could be further combined with the loading-unloading model (Di Luca et al. 2011) by an appropriate weighting of the data points in calculation of the regression parameters.

Certainly, other models could be devised based on additional studies. For example, in the attempt to augment the perceived stiffness of virtual surfaces with simple haptic devices that cannot generate high forces, models based on Rate Hardness (Han and Choi 2010; Lawrence et al. 2000), i.e. the ratio between change of force and velocity at specific points of interaction with the surface, were suggested. According to the Rate Hardness model (Lawrence et al. 2000), perceptual judgment of rigidity can be explained as the ratio between initial change in force and initial velocity, where initial means at the instance when the user first encounters the virtual surface. In the Extended Rate Hardness model (Han and Choi 2010), the initial force change is replaced with maximal force change. A model for perception of stiffness with delay could be constructed such that it takes into account rate hardness. Naturally, some adjustment might be necessary for the model to explain as much as possible of the answers of our participants. For example, in the cases when participants choose not to cross the boundary of the elastic force field while probing (Nisky et al. 2008b), or could not access it (Nisky et al. 2010), the initial interaction with the elastic force field is not well defined, and therefore, maximum velocity could be used instead (Han and Choi 2010).

We have made the experimental data from several of our studies reviewed in the current chapter available in an online repository located at http://www.bgu.ac.il/~akarniel/database/index.htm. Hopefully, new models for perception of stiffness will be tested against the answers of participants in our experiments. We invite the readers to contact us if they wish to add their experimental data to this repository.

9.4.2 Development of Novel Solutions for Teleoperation

Future experiments will be necessary to elucidate the underlying mechanisms that are responsible for the behavioral effects, and it is very likely that the mechanistic aspects of the models will be refined or completely changed in the process. The empirical predictions of the models may be well used in practical applications such as improving transparency in teleoperation. Transparency typically refers to the ability of a teleoperation system to accurately display the mechanical properties of the remote environment to the operator, and to accurately execute operator's actions in the remote environment. While this intuitive definition includes the operator and the environment, most of the classical studies of transparency focused on the optimization of teleoperation system in isolation from the human operator and the environment—a system-centered approach. Recently, an alternative, human-centered, approach to this problem was suggested (Nisky 2011; Nisky et al. 2008a, 2011, 2013). A key step in developing such a human-centered approach to remote teleoperation was to explore the effects of delay on the perception and action of the operator.

Based on the findings that we discussed in the current chapter, a multidimensional measure of transparency in teleoperation was suggested, which included the following components:

1. *Perceptual transparency*: The human operator cannot distinguish between teleoperation with the system and direct operation.
2. *Local motor transparency*: The movement of the operator (position and force trajectories) does not change when the teleoperation system is replaced with direct operation at local site.
3. *Remote motor transparency*: The movement of the remote robot (position and force trajectories) does not change when the teleoperation system is replaced with direct operation at remote site.

According to this framework, in non-ideal systems, it is beneficial to obtain perceptually transparent teleoperation (1) and remote motor transparency (3) without local motor transparency (2). The gap between perception and action that we discussed is a necessary condition for transparency in such non-ideal system, because a choice must be made between local and remote motor transparencies. In practice, local motor transparency is relatively unimportant, since the motor goal of teleoperation tasks is often defined in the remote environment. Therefore, we can sacrifice local motor transparency in favor of remote motor transparency. At the same time, however, realistic perception must be rendered in the local environment. The gap between perception and action can allow perceptual transparency without the motor counterpart at the local site.

In addition, in cases where the baseline gap between perception and action is insufficient for allowing perceptual as well as remote motor transparency optimization, training of the human operator can be an effective strategy to improve transparency. Such training-induced dissociation between perception and action was reported in many studies of motor adaptation, including adaptation to force fields (Shadmehr and Mussa-Ivaldi 1994), and grip force adaptation in face of the size-weight illusion (Brayanov and Smith 2010; Flanagan and Beltzner 2000).

This strategy was successfully employed to achieve transparency in a virtual one-dimensional remote needle insertion task (Nisky et al. 2011). The virtual teleoperation channel consisted from a pure delay and a gain, and the environment was a nonlinear, needle insertion-like force field: a nonlinear rigid boundary that represents penetration into the skin, followed by a linear elastic force field representing the underlying soft tissue. Using the protocol described in the current chapter, the authors experimentally identified the effect of delay on the answers of participants about their stiffness perception and on their probability to overshoot the inserted needle. They found that the motor response of the participants indicated underestimation of the stiffness of the nonlinear elastic field when compared to its non-delayed counterpart, but the perception of participants was not biased. Based on these findings, the authors calculated the gains that achieved perceptual and motor transparency, and demonstrated the applicability of the new empirical approach for transparency optimization.

9.5 Conclusions

In this chapter, we reviewed how external delay between probing movements and contact forces changes perception of mechanical stiffness. The effect of delay on perception is complex, and depends on many factors such as the way the user interacts with the probed objects, the joint that is used during probing, the source of information that is delayed, and possibly, even the range of stiffness levels of the probed object. Moreover, there is a gap between the effect of delay on cognitive perception of mechanical properties of objects and its effect on action in contact with them.

Understanding delay effects is important because the temporal mismatch between force and movement signals decouples them, and provides a window into the sensorimotor processes underlying perception of mechanical properties. In addition, these studies can also teach us about how our brain copes with internal delays. Finally, it is important because it can provide the necessary information for human-centered design and control of teleoperation systems. We accompany the chapter with an online repository that contains raw data from several of our studies that were reviewed in this chapter. We hope that the repository will be used to further promote each of these important goals.

Acknowledgments This work was funded by Grant No. 2003021 from the United States-Israel Binational Science Foundation (BSF), Jerusalem, Israel, and by the Israeli Science Foundation (ISF), Jerusalem, Israel. IN was supported by the Weizmann Institute National Program for Promoting Women in Science, and by the Marie Curie International Outgoing Fellowship.

References

Aglioti S, DeSouza JFX, Goodale MA (1995) Size-contrast illusions deceive the eye but not the hand. Curr Biol 5(6):679–685

Anvari M (2007) Remote telepresence surgery: the Canadian experience. Surg Endosc 21(4):537–541

Avraham G, Nisky I, Fernandes H, Acuna D, Kording K, Loeb G, Karniel A (2012) Towards perceiving robots as humans–three handshake models face the Turing-like handshake Test. IEEE Trans Haptics 5(3):196–207

Ballantyne GH (2002) Robotic surgery, telerobotic surgery, telepresence, and telementoring. Surg Endosc 16(10):1389–1402. doi:10.1007/s00464-001-8283-7

Bicchi A, Scilingo EP, Ricciardi E, Pietrini P (2008) Tactile flow explains haptic counterparts of common visual illusions. Brain Res Bull 75(6):737–741

Biggs J, Srinivasan MA (2002) Haptic interfaces. In: Stanney K (ed) Handbook of virtual environments. Lawrence Earlbaum Inc, London, pp 93–115

Box GEP, Draper NR (1987) Empirical model-building and responce surfaces. Wiley, New Jersey

Brayanov JB, Smith MA (2010) Bayesian and "Anti-Bayesian" biases in sensory integration for action and perception in the size-weight illusion. J Neurophysiol 103(3):1518–1531. doi:10.1152/jn.00814.2009

Brouwer B, Ashby P (1990) Corticospinal projections to upper and lower limb spinal motoneurons in man. Electroencephalogr Clin Neurophysiol 76(6):509–519

Brown J, Rosen J, Chang L, Sinanan M, Hannaford B (2004) Quantifying surgeon grasping mechanics in laparoscopy using the blue DRAGON system. Paper presented at the Medicine Meets Virtual Reality, Newport Beach, CA

Carey DP (2001) Do action systems resist visual illusions? Trends Cogn Sci 5(3):109–113

Chib VS, Krutky MA, Lynch KM, Mussa-Ivaldi FA (2009) The separate neural control of hand movements and contact forces. J Neurosci 29(12):3939–3947

Chib VS, Patton JL, Lynch KM, Mussa-Ivaldi FA (2006) Haptic identification of surfaces as fields of force. J Neurophysiol 95(2):1068–1077

Colebatch JG, Gandevia SC (1989) The distribution of muscular weakness in upper motor neuron lesions affecting the arm. Brain 112:749–763

Cooper JR, Wernke MM, Reed KB (2012) The effects of incongruent feedback on bimanual task performance. Paper presented at the Haptics symposium (HAPTICS), IEEE, 4–7 March 2012

Davidson A, Buford J (2006) Bilateral actions of the reticulospinal tract on arm and shoulder muscles in the monkey: stimulus triggered averaging. Exp Brain Res 173(1):25–39

De Gersem G, Van Brussel H, Tendick F (2005) Reliable and enhanced stiffness perception in soft-tissue telemanipulation. Int J Robot Res 24(10):805–822. doi:10.1177/0278364905057861

Di Luca M, Knorlein B, Ernst MO, Harders M (2011) Effects of visual-haptic asynchronies and loading-unloading movements on compliance perception. Brain Res Bull 85(5):245–259

Dijkerman HC, Vargha-Khadem F, Polkey CE, Weiskrantz L (2008) Ipsilesional and contralesional sensorimotor function after hemispherectomy: differences between distal and proximal function. Neuropsychologia 46(3):886–901

Domenico G, McCloskey DI (1987) Accuracy of voluntary movements at the thumb and elbow joints. Exp Brain Res 65(2):471–478

Duong MD, Terashima K, Miyoshi T, Okada T (2010) Rehabilitation system using teleoperation with force-feedback-based impedance adjustment and EMG-moment model for arm muscle strength assessment. J Rob Mechatron 22(1):10–20

Ernst MO, Banks MS (2002) Humans integrate visual and haptic information in a statistically optimal fashion. Nature 415(6870):429–433

Fernandes HL, Kording KP (2010) In praise of "False" models and rich data. J Mot Behav 42(6):343–349. doi:10.1080/00222895.2010.526462

Flanagan JR, Beltzner MA (2000) Independence of perceptual and sensorimotor predictions in the size-weight illusion. Nat Neurosci 3(7):737–741

Ganel T, Goodale MA (2003) Visual control of action but not perception requires analytical processing of object shape. Nature 426(6967):664–667

Gentaz E, Hatwell Y (2004) Geometrical haptic illusions: the role of exploration in the Müller-Lyer, vertical-horizontal, and Delboeuf illusions. Psychon Bull Rev 11(1):31–40. doi:10.3758/bf03206457

Gillespie R, O'Modhrain M, Tang P, Zaretzky D, Pham C (1997) The virtual teacher. Paper presented at the ASME dynamic systems and control division, Anaheim, CA

Goodale MA, Humphrey GK (1998) The objects of action and perception. Cognition 67(1–2):181–207

Goodale MA, Milner AD (1992) Separate visoual pathways for perception and action. Trends Neurosci 15(1):20–25

Hall LA, McCloskey DI (1983) Detections of movements imposed on finger, elbow and shoulder joints. J Physiol 335(1):519–533

Hamilton AF, Jones KE, Wolpert DM (2004) The scaling of motor noise with muscle strength and motor unit number in humans. Exp Brain Res 157(4):417–430

Han G, Choi S (2010) Extended rate-hardness: a measure for perceived hardness. In: Kappers AML, Van Erp JF, Bergmann Tiest WM, Van der Helm FCT (eds) Haptics: generating and perceiving tangible sensations, vol 6191. Springer, Berlin Heidelberg, pp 117–124

Hirche S, Bauer A, Buss M (2005) Transparency of haptic telepresence systems with constant time delay. Paper presented at the proceedings of the 2005 IEEE conference on control applications, Toronto, Canada

Hirche S, Buss M (2007) Human perceived transparency with time delay. Adv Telerobotics 191–209

Hirzinger G, Brunner B, Dietrich J, Heindl J (1993) Sensor-based space robotics: ROTEX and its telerobotic features. IEEE Trans Robot Autom 9(5):649–663

Hogan N, Kay BA, Fasse ED, Mussa-Ivaldi FA (1990) Haptic illusions: experiments on human manipulation and perception of "Virtual Objects". Cold spring Harbor symposia on quantitative biology 55:925–931. doi:10.1101/sqb.1990.055.01.086

Imaida T, Yokokohji Y, Doi T, Oda M, Yoshikawa T (2004) Ground-space bilateral teleoperation of ETS-VII robot arm by direct bilateral coupling under 7-s time delay condition. IEEE Trans Robot Autom 20(3):499–511

Ishihara M, Negishi N (2008) Effect of feedback force delays on the operation of haptic displays. IEEJ Trans Electr Electron Eng 3(1):151–153

Kaim L, Drewing K (2008) Exploratory movement parameters vary with stimulus stiffness. In: Ferre M (ed) Lecture notes in computer science, vol 5024. Springer, Berlin / Heidelberg, pp 313–318

Kandel E, Schwartz J, Jessel T (2000) Principles of neural science, 4th edn. McGraw-Hill

Karniel A (2009) Computational motor control. In: Binder M, Hirokawa N, Windhorst U (eds) Encyclopedic reference of neuroscience. Springer, Berlin Heidelberg, pp 832–837

Karniel A (2011) Open questions in computational motor control. J Integr Neurosci 10:385–411

Karniel A, Nisky I, Avraham G, Peles B-C, Levy-Tzedek S (2010) A turing-like handshake test for motor intelligence. In: Kappers AML, Van Erp JF, Bergmann Tiest WM, Van der Helm FCT (eds) Haptics: generating and perceiving tangible sensations, vol 6191. Springer, Berlin/Heidelberg, pp 197–204

Kawato M (1999) Internal models for motor control and trajectory planning. Curr Opin Neurobiol 9:718–727

Kurata K, Tanji J (1986) Premotor cortex neurons in macaques: activity before distal and proximal forelimb movements. J Neurosci 6(2):403–411

Lawrence DA, Pao LY, Dougherty AM, Salada MA, Pavlou Y (2000) Rate-hardness: a new performance metric for haptic interfaces. IEEE Trans Robot Autom 16(4):357–371

Lederman SJ, Jones LA (2011) Tactile and haptic illusions. IEEE Trans Haptics 4(4):273–294. doi:10.1109/toh.2011.2

Leib R, Nisky I, Karniel A (2010) Perception of stiffness during interaction with delay-like nonlinear force field. Paper presented at the EuroHaptics 2010, part 1, LCNS 6191, Amsterdam

Lemon RN, Griffiths J (2005) Comparing the function of the corticospinal system in different species: organizational differences for motor specialization? Muscle Nerve 32(3):261–279

Lu LH, Barrett AM, Cibula JE, Gilmore RL, Heilman KM (2000) Proprioception more impaired distally than proximally in subjects with hemispheric dysfunction. Neurology 55(4):596–597

Marescaux J, Leroy J, Gagner M, Rubino F, Mutter D, Vix M, Smith MK (2001) Transatlantic robot-assisted telesurgery. Nature 413(6854):379–380

McKiernan BJ, Marcario JK, Karrer JH, Cheney PD (1998) Corticomotoneuronal postspike effects in shoulder, elbow, wrist, digit, and intrinsic hand muscles during a reach and prehension task. J Neurophysiol 80(4):1961–1980

Millman PA, Colgate JE (1995) Effects of nonuniform environment damping on haptic perception and performance of aimed movements. Paper presented at the ASME Dynamic Systems and Control Division

Mugge W, Schuurmans J, Schouten AC, van der Helm FCT (2009) Sensory weighting of force and position feedback in human motor control tasks. J Neurosci 29(17):5476–5482

Nisky I (2011) Perceptuomotor transparency in bilateral teleoperation: the effect of delay on perception and action. Ph.D. thesis, Ben-Gurion University of the Negev, Beer-Sheva, Israel

Nisky I, Baraduc P, Karniel A (2010) Proximodistal gradient in the perception of delayed stiffness. J Neurophysiol 103(6):3017–3026

Nisky I, Mussa-Ivaldi FA, Karniel A (2008a) Perceptuo-motor transparency in bilateral teleoperation. In: ASME conference proceedings, pp 449–456

Nisky I, Mussa-Ivaldi FA, Karniel A (2008b) A regression and boundary-crossing-based model for the perception of delayed stiffness. IEEE Trans Haptics 1(2):73–82

Nisky I, Mussa-Ivaldi FA, Karniel A (2013) Analytical study of perceptual and motor transparency in bilateral teleoperation. IEEE Trans Hum Mach Syst 43(6):570–582

Nisky I, Pressman A, Pugh CM, Mussa-Ivaldi FA, Karniel A (2010) Perception and action in simulated telesurgery. Paper presented at the EuroHaptics 2010, Amsterdam, Netherlands

Nisky I, Pressman A, Pugh CM, Mussa-Ivaldi FA, Karniel A (2011) Perception and action in teleoperated needle insertion. IEEE Trans Haptics 4(3):155–166

Ohnishi H, Mochizuki K (2007) Effect of delay of feedback force on perception of elastic force: a psychophysical approach. IEICE Trans Commun E-B 90(1):12–20

Okamoto S, Konyo M, Saga S, Tadokoro S (2009) Detectability and perceptual consequences of delayed feedback in a vibrotactile texture display. IEEE Trans Haptics 2(2):73–84. doi:10.1109/toh.2009.17

Palmer E, Ashby P (1992) Corticospinal projections to upper limb motoneurones in humans. J Physiol 448(1):397–412

Pressman A, Karniel A, Mussa-Ivaldi FA (2011) How soft is that pillow? The perceptual localization of the hand and the haptic assessment of contact rigidity. J Neurosci 31(17):6595–6604. doi:10.1523/jneurosci.4656-10.2011

Pressman A, Nisky I, Karniel A, Mussa-Ivaldi FA (2008) Probing virtual boundaries and the perception of delayed stiffness. Adv Rob 22:119–140

Pressman A, Welty LJ, Karniel A, Mussa-Ivaldi FA (2007) Perception of delayed stiffness. Int J Robot Res 26(11–12):1191–1203

Raibert MH, Craig JJ (1981) Hybrid position/force control of manipulators. ASME J Dyn Syst Meas Contr 102(2):126–133

Refshauge KM, Chan R, Taylor JL, McCloskey DI (1995) Detection of movements imposed on human hip, knee, ankle and toe joints. J Physiol 488(1):231–241

Reinkensmeyer DJ, Pang CT, Nessler JA, Painter CC (2002) Web-based telerehabilitation for the upper extremity after stroke. IEEE Trans Neural Syst Rehabil Eng 10(2):102–108

Reintsema D, Landzettel K, Hirzinger G (2007) DLR's advanced telerobotic concepts and experiments for on-orbit servicing. In: Advances in telerobotics, pp 323–345

Repperger DW, Phillips CA, Chelette TL (1995) A study on spatially induced "virtual force" with an information theoretic investigation of human performance. IEEE Trans Syst Man Cybern 25(10):1392–1404

Riddle CN, Edgley SA, Baker SN (2009) Direct and indirect connections with upper limb motoneurons from the primate reticulospinal tract. J Neurosci 29(15):4993–4999

Rosser JC, Wood M, Payne JH, Fullum TM, Lisehora GB, Rosser LE, Savalgi RS (1997) Telementoring. Surg Endosc 11(8):852–855. doi:10.1007/s004649900471

Satava RM (2006) Robotics in colorectal surgery: telemonitoring and telerobotics. Surg Clin North Am 86(4):927–936

Shadmehr R, Mussa-Ivaldi FA (1994) Adaptive representation of dynamics during learning of a motor task. J Neurosci 14(5):3208–3224

Sheridan TB (1997) Eight ultimate challenges of human-robot communication. Paper presented at the RO-MAN '97, IEEE international workshop on robot and human communication

Tan HZ, Srinivassan MA, Eberman B, Cheng B (1994) Human factors for the design of force reflecting haptic interfaces. In: Radcliffe CJ (ed) ASME DSC dynamic systems and control, vol 55–1, pp 353–359

Turton A, Lemon RN (1999) The contribution of fast corticospinal input to the voluntary activation of proximal muscles in normal subjects and in stroke patients. Exp Brain Res 129(4):559–572

Venkadesan M, Valero-Cuevas FJ (2008) Neural control of motion-to-force transitions with the fingertip. J Neurosci 28(6):1366–1373

Wichmann F, Hill N (2001) The psychometric function: I. Fitting, sampling, and goodness of fit. Percept Psychophysics 63(8):1293–1313

Wolpert DM, Kawato M (1998) Multiple paired forward and inverse models for motor control. Neural Netw 11(7–8):1317–1329

Wolpert DM, Miall RC, Kawato M (1998) Internal models in the cerebellum. Trends Cogn Sci 2(9):338–347

Yip MC, Tavakoli M, Howe RD (2011) Performance analysis of a haptic telemanipulation task under time delay. Adv Rob 25(5):651–673

Yoon W, Goshozono T, Kawabe H, Kinami M, Tsumaki Y, Uchiyama M, Doi T (2004) Model-based space robot teleoperation of ETS-VII manipulator. IEEE Trans Rob Autom 20(3):602–612

Part III
Artificial Softness

Chapter 10
Compliance Perception Using Natural and Artificial Motion Cues

Netta Gurari and Allison M. Okamura

10.1 Introduction

In order to perform activities such as pressing keys on a keyboard, checking whether a piece of fruit is ripe, determining whether a bike tire is low on air, and shaking another person's hand, it is necessary to have an understanding of the object's compliance, or the relationship between one's applied force and the resulting change in position of one's hand. There are many situations in which humans are not able to perform such tasks due to limitations in their sensory and/or motor control channels. For example, upper-limb prosthesis users lack perception of sensations at the artificial limb, and surgeons performing procedures using teleoperated robots lack haptic feedback. For those with suboptimal sensing, the quality of experiencing interactions with the environment and the ability to perform manual tasks may be enhanced by the development of new technologies and paradigms that will artificially create or enhance missing sensations.

When performing activities of daily living, humans must merge kinesthetic and tactile cues (Bicchi et al. 2000; Srinivasan and LaMotte 1995) to sense an objects compliance. This means that they combine position information derived from sensory receptors in muscles, articular capsulae, and tendons (kinesthetic cues) with pressure and indentation information derived from the skin (tactile cues). Kinesthetic cues, collectively referred to as kinaesthesia, were first defined by Bastian (1887), and the term has been used synonymously with the term proprioception later defined by Burke (2007). Here we refer to proprioception as the intuitive knowledge for how one's body is situated in space and how it moves (in the absence of vision or other external sensory signals). A distinction can be made between kinaesthesia and proprioception, where

N. Gurari (✉)
Department of Robotics, Brain and Cognitive Sciences (RBCS), Istituto Italiano di Tecnologia, Via Morego 30, 16123 Genoa, Italy

N. Gurari
Faculty of Psychology, San Raffaele Vita-Salute University, Milan, Italy
e-mail: netta.gurari@iit.it

A.M. Okamura
Stanford University, 416 Escondido Mall, Stanford, CA 94305, USA
e-mail: aokamura@stanford.edu

© Springer-Verlag London 2014
M. Di Luca (ed.), *Multisensory Softness*, Springer Series on Touch and Haptic Systems,
DOI 10.1007/978-1-4471-6533-0_10

the former refers to limb movement and the latter to limb position; however, here we use the two terms interchangeably. Prior work demonstrates the importance of tactile cues for interpreting compliance, especially when interacting with deformable objects such as a piece of rubber or fruit (Srinivasan and LaMotte 1995). Here, we focus on compliance perception when interacting with rigid-surfaced objects, such as piano or keyboard keys, where both kinesthetic and tactile cues are needed. We hypothesize that proprioception plays an important role for interacting with such rigid-surfaced compliant objects and that it should be considered when designing systems for artificially relaying compliance.

Ian Waterman is a famous patient who lost his proprioception when he was afflicted by a large-fiber sensory neuropathy (Cole 1995). Ian is unique regarding his neuropathy, in that he learned to control his movements and interact with the world again through a reliance on sight. If the lights are turned off in a room or if he is distracted by an attractive woman, he loses control and collapses. Ian describes the energy he exerts to control his motions on a regular basis as running a daily marathon. Thus, the lack of one's sensing capability, notably proprioception, results in a heavy taxing of the visual sensory modality, and further, daily living activities are not possible for most individuals afflicted by such a neuropathy.

In this chapter, we present challenges faced in the design of active touch feedback devices for the artificial display of compliance (e.g. using actuators in a tactile feedback system to stimulate, and in turn, convey information to the user). We hypothesize that relaying proprioception will provide significant benefit to the user. Below we describe related literature that discusses the role of proprioception in compliance perception. Then, we present a human participant study, conducted using a novel setup that allows the role of motion cues to be quantified in a compliance perception task under differing sensory conditions. Next, we discuss methods by which compliance can be relayed artificially using active haptic feedback devices, and present the advantages and disadvantages of various touch sensory substitution systems (i.e., electrocutaneous, vibrotactile, force, and skin stretch stimulation). Last, we discuss two human studies in which the effectiveness of a skin stretch feedback system and a vibrotactile feedback system is tested for displaying respectively compliance and proprioception information. The skin stretch feedback device was tested for numerous reasons, including that it has been shown to be more effective in relaying positional information than the more traditional vibratory feedback devices and it allows direct mapping between a rotation of one's limb and a rotation of the skin stretch device. The chapter concludes with a summary of what has been presented, and a proposal for future research directions.

10.2 The Role of Proprioception in Human Compliance Perception

10.2.1 Background

The mechanism for human compliance perception is complex and is still not completely understood. An object's compliance can be estimated by merging the aforementioned proprioceptive, force, and tactile signals (Kuschel et al. 2010; Pressman et al. 2007; Srinivasan et al. 1996; Srinivasan and LaMotte 1995; Tiest and Kappers 2009; Varadharajan et al. 2008). Interestingly, visual and auditory information also influence one's estimation of an object's haptic compliance (Avanzini and Crosato 2006; Lecuyer et al. 2000); these phenomena are discussed in detail in separate chapters. As humans do not have dedicated compliance mechanoreceptors for directly measuring an object's elasticity, a common hypothesis is that compliance perception occurs by combining force, position, and cutaneous cues. The manner in which such cues combine to identify an object's compliance has been hypothesized and tested, e.g. (Pressman et al. 2007; Tiest and Kappers 2009). Here, we focus on the role of motion information (proprioception) in compliance perception.

Proprioception plays an important role in object interaction tasks, as it provides a context with which other touch cues can integrate in order for a person to understand his/her surroundings (Berryman et al. 2006). As with compliance, the mechanism for how proprioception occurs is not known; that is, the manner by which the various signals are combined and/or integrated remains unclear (Matthews 1982). It is known that proprioception is derived from numerous sources, including muscle spindle fibers, Golgi tendon organs, joint angle receptors, and skin stretch mechanoreceptors (Collins and Prochazka 1996; Collins et al. 2005; Edin 2004; McCloskey 1978). Additionally, corollary discharges, or copies of the efferent commands that are sent to the sensory area of the brain, can create the perception of limb motion even in the absence of limb motion (Crapse and Sommer 2008; Gandevia et al. 2006). The amount each signal contributes to the sensation may differ depending on which limb is being rotated and the movement type that is being made, e.g. (Brewer et al. 2005) versus (Tan et al. 2007), as well as the level of muscle activation (Taylor and McCloskey 1992).

Given the complexity of proprioception, it is difficult to remove this sensation without affecting other sensory capabilities (e.g. cutaneous cues), making it very challenging to study the role of proprioception in sensorimotor tasks. Interestingly, there are a small number of individuals afflicted by large-fiber sensory neuropathies. They do not have proprioception, but have intact cutaneous sensations. It is possible to compare task performance between these individuals and healthy participants in the presence and absence of proprioception (Ghez et al. 1990). This method, however, does not allow for comparison of performance with and without proprioception in a single person; thus, the role of proprioception in the task might be confounded by other unanticipated or unacknowledged factors that are occurring between the persons. Positional cues can be removed using invasive methods [i.e., anesthesia (McCloskey 1978) or ischemic nerve blocks (Kelso 1977)]. With such interventions, the role of tactile cues and proprioception cannot be disentangled, and there are practical limitations due to the invasiveness of the procedures. The position information can be non-invasively altered by vibrating particular muscles (Larish et al. 1984), but the sensation is shifted and not removed. Additionally, the role of motion cues has been evaluated by either permitting or restricting the motion of a limb in the presence of sight (Blank et al. 2010). When the limb motion is constrained, proprioceptive cues from the constrained limb indicate that the limb is not moving, while the visual cues indicate that the limb is moving. Thus, sensory integration may be occurring such that the "visual motion" cues are based on both the moving visual and non-moving proprioceptive information.

10.2.2 Experiment: Role of Motion Cues in Compliance Perception

Here we present an experiment that investigates how motion cues (i.e., visual, proprioceptive) and forces cues combine and how visual and proprioceptive motion cues integrate when pressing a compliant object. More details about the study can be found in Gurari et al. (2013). The motivation for this work was to investigate whether the upper-limb prosthesis experience can be improved by reducing the demand on vision during arm control. The power of visual cues for eliciting haptic sensations is demonstrated in a separate chapter; thus, we do not delve into a discussion of this literature here.

For upper-limb prosthesis users, it has been found that a phantom limb can be moved just by observing a visual limb's motion that is initially co-located with the phantom limb (Ramachandran and Rogers-Ramachandran 1996; Ramachandran et al. 1995). Upper-limb prosthesis users, however, desire their artificial limbs to require less visual attention during certain tasks (Atkins et al. 1996). Thus, we aim to quantify the effectiveness of various sensory modalities for relaying positional cues, with the eventual goal of artificially relaying such information to those with compromised sensing.

15 healthy, intact, dominantly right-handed participants with no neurological illnesses or right-hand impairments took part in the study. Eight participants were male and seven female, with an age range of 18–34. The Johns Hopkins University Homewood Institutional Review Board provided approval to run this study, and all participants provided their informed consent before taking part.

Participants used a custom kinesthetic feedback haptic device to control the motion of either their real finger (proprioceptively) or of a virtual finger displayed on a computer monitor (visually) for a one-degree-of-freedom rotational spring discrimination task by applying a force at their index finger. Thus, participants always had corollary discharge cues since they were actively applying an effort, and they always felt the sensation of an applied force at their finger. The manner by which the motion was portrayed was either using visual motion cues, proprioceptive motion cues, or their combination. Note that there is always proprioception; the distinction between the conditions is whether the proprioceptive cues indicate that the finger is in motion or that the finger is held still. Performance for a spring discrimination task was tested under each of these conditions using the method of constant stimuli for a two-alternative forced choice task on rigid-surfaced virtual springs, with a reference stiffness of 290 N/m.

Below we briefly summarize the experiment, discuss the general findings, and explain the implications for how sensory combination and integration occur. Additionally, we motivate the need for either creating or enhancing the motion sensation for those with compromised proprioception using active artificial feedback devices. Last, we investigate whether the (Ernst and Banks 2002) maximum-likelihood estimation (MLE) model, which states that visual and proprioceptive information combine according to the MLE theory during a compliance discrimination task, can explain the sensory integration of visual and proprioceptive cues for the task discussed here.

10.2.2.1 Sensory Conditions

A novel testbed was created with the aim of investigating the role of visual motion and/or proprioceptive motion cues in the perception of an object's compliance (see Fig. 10.1). The role of vision is relatively straightforward to study since a cue can be either displayed or hidden from view on a computer monitor. The role of proprioception was investigated by either allowing the limb to move (isotonic condition), or by restraining its motion (isometric condition).

The mechanism by which compliance may be sensed in our study is visually depicted in Fig. 10.2. Throughout the experiment, participants always applied a force in order to control the compression of a virtual spring. The manner by which the motion cues were rendered was varied as follows: visually a virtual finger moved on a computer monitor (*Visual Motion*), physically the real finger rotated (*Proprioceptive Motion*), or a combination of the two occurred (*Visual and Proprioceptive Motion*). In all conditions, sensory combination occurs with corollary discharge cues, cutaneous/force/torque cues, and position/motion cues contributing to the compliance percept. The manner by which the position/motion cues contribute differs depending on which sensory modalities are relaying the sensation—vision and/or proprioception—and how they integrate.

During the *Visual Motion* condition, sensory integration may occur since a virtual finger is visually drawn to show the motion of the finger and the real finger proprioceptively indicates that the finger is not moving. Note that this is similar to what may occur during upper-limb prosthesis use; the visual cues of the artificial limb movement may integrate with the proprioceptive phantom limb cues of the "stationary arm", as in the rubber hand illusion (Ramachandran et al. 1995). We also point out that the virtual finger is not an accurate representation of a real finger. Likewise, a

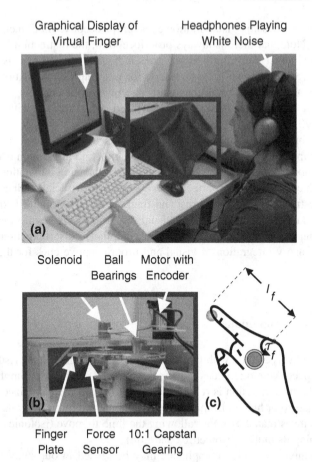

Fig. 10.1 **a** Experimental setup. A computer monitor displays a graphical depiction of the user's index finger, and white noise playing headphones mask possible auditory cues/distractions. The user interacts with the haptic device using the right index finger (*blue box*), which is concealed from sight by the red cloth. **b** Custom haptic device. The ball bearings permit rotation of the finger plate, and the motor with encoder applies a torque and measures the angular position of the finger plate via a 10:1 capstan gear. The force sensor measures the user's applied effort, and the solenoid can lock the finger plate in a stationary position. **c** User interaction. The user applies a torque about the metacarpophalangeal joint of τ_f, has a right index finger length of l_f, and makes contact with the finger plate at the volar portion of the right index finger, as indicated by the orange circle. Adapted from Fig. 10.1 in Gurari et al. (2013) © IEEE 2013

prosthetic limb may not look like a natural limb. During the *Proprioceptive Motion* condition, sensory integration does not occur since only haptic cues are available. During the *Visual and Proprioceptive Motion* condition, sensory integration may occur since both the visually relayed and proprioceptively sensed motion cues may be used in determining the sensation of motion.

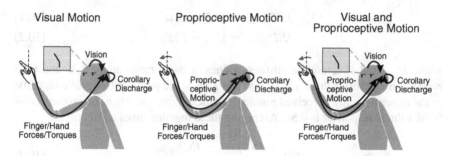

Fig. 10.2 Experimentally tested sensory conditions. Adapted from Fig. 10.2 in Gurari et al. (2013) © IEEE 2013

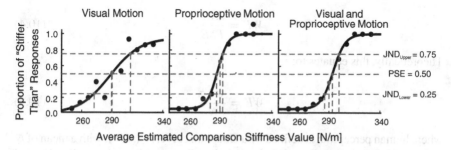

Fig. 10.3 Psychometric curves based on experimental data for an example participant. Raw data is indicated by the *black dots*, nominal stiffness ($N = 290$ N/m) by the *green line*, *PSE* by the *red line*, and *lower* and *upper JND*s by the *blue lines*. Adapted from Fig. 10.3 in Gurari et al. (2013) © IEEE 2013

10.2.2.2   Results and Discussion

Participants interacted with 10 comparison springs (5 equally-spaced springs with a stiffness that is less than 290 N/m and 5 with a stiffness that is greater than 290 N/m) and indicated whether the comparison spring was stiffer than the nominal spring (stiffness of 290 N/m). The proportion of "stiffer than" responses is plotted as a function of the comparison spring stiffness value, giving psychometric curves as displayed in Fig. 10.3 (example curves for a representative participant). Task performance was quantified in each condition using the Weber fraction (WF), which relates a physical change to how one perceives the change (Gescheider 1997). WFs were estimated based on the psychometric curves that were experimentally obtained for each participant and every condition.

First, the just noticeable difference (JND), or amount of change in stiffness that was required for the change to be noticed, was estimated by identifying the lower and upper JND:

$$JND_{lower} = PSE - k^*_{0.25} \qquad (10.1)$$

$$JND_{upper} = k^*_{0.75} - PSE \qquad (10.2)$$

where $k^*_{0.25}$ and $k^*_{0.75}$ are the stiffness values at the corresponding 0.25 and 0.75 proportion of "stiffer than" responses, and PSE is the point-of-subjective-equality, or the experimentally perceived nominal stiffness value, N, when the proportion of "stiffer than" responses is 0.50. Averaging these together gives a JND of:

$$JND = \frac{JND_{lower} + JND_{upper}}{2}. \qquad (10.3)$$

In turn, the WF was experimentally obtained as follows:

$$WF = \frac{JND}{PSE}. \qquad (10.4)$$

Theoretically, this equates to:

$$WF = \frac{\sqrt{2}\sigma}{N}, \qquad (10.5)$$

where human perception is defined by a Gaussian curve, $\mathcal{N}(N, \sigma)$, with a mean of N and standard deviation of σ. $JND = \sqrt{2}\sigma$ describes one's experimentally obtained perceptual capabilities when pressing two springs, assuming that equal levels of noise, σ, are sensed when pressing each spring.

Ideally, the experimentally obtained PSE values across all tested conditions are 290 N/m, but in practice they have differing values that depend on the noise in the sensing and rendering capabilities. Thus, it made sense to compare the unit-less WFs rather than the JNDs with units of Newtons, where the noise level was normalized by the experimentally obtained nominal stiffness values. A lower WF indicates better perceptual/sensing capabilities.

Task performance, as defined by the WFs, and user ratings were used to compare compliance perception under *Visual Motion*, *Proprioceptive Motion*, and *Visual and Proprioceptive Motion* cues. The main findings from this work are as follows:

- no significant differences were found in task performance among the three sensory conditions,
- the MLE model was not able to explain the signal dispersion observed in the *Visual and Proprioceptive Motion* condition based on task performance in the *Visual Motion* and *Proprioceptive Motion* conditions,
- exploration style (largest penetration distance, number of spring presses, and total exploration time) did not significantly differ between sensory conditions, and
- the *Proprioceptive Motion* and *Visual and Proprioceptive Motion* conditions were subjectively rated by users as significantly more useful than *Visual Motion*.

Task Performance

The spring discrimination sensory integration literature supports the notion that vision dominates when users perceive a spring, and that the addition of vision to proprioception improves perceptual performance. Paljic et al. (2004) ran a spring discrimination study in which participants discriminated between isometric and elastic haptic springs in the presence of vision. During the isometric condition the user applies a force yet the finger does not move, while during the elastic condition the user applies a force and the finger rotates an amount proportional to the measured force. They found that the visual feedback can create a motion percept for the isometric condition, and the combination of the proprioceptive and visual motion cues in the elastic condition gives improved perceptual performance over the isometric condition. Based on anecdotal feedback from our participants, we also observed visual dominance in the *Visual Motion* condition; several participants commented that they felt that their hand was moving. Unlike (Paljic et al. 2004), compliance discrimination performance was not found to differ across the three conditions of *Visual Motion*, *Proprioceptive Motion*, and *Visual and Proprioceptive Motion* (see Table 10.1 for an overview of the WF results). A within-subjects one-way analysis of variance (ANOVA) with a Box (Greenhouse-Geisser) epsilon-hat adjustment did not find a statistically significant difference between the conditions [$F(2, 22) = 2.21$, $\hat{\varepsilon} = 0.6041, p = 0.16$].

MLE Model

We tested whether the Ernst and Banks MLE model, which states that visual cues, $\mathcal{N}\,(PSE_v, \sigma_v)$, and proprioceptive cues, $\mathcal{N}\,(PSE_p, \sigma_p)$, integrate in a compliance discrimination task according to the maximum-likelihood estimation (MLE), holds (Ernst and Banks 2002):

$$PSE_{MLE} = \frac{\sigma_p^2}{\sigma_v^2 + \sigma_p^2} PSE_v + \frac{\sigma_v^2}{\sigma_v^2 + \sigma_p^2} PSE_p, \tag{10.6}$$

$$\sigma_{MLE} = \sqrt{\frac{\sigma_v^2 \sigma_p^2}{\sigma_v^2 + \sigma_p^2}}, \tag{10.7}$$

and

$$WF_{MLE} = \frac{\sqrt{2}\,\sigma_{MLE}}{PSE_{MLE}}. \tag{10.8}$$

Human sensory perception of a stimulus is modeled by a Gaussian curve with a signal, PSE_{MLE}, and signal dispersion, σ_{MLE}. These parameters are used to calculate a theoretical Weber fraction (WF_{MLE}).

We used our results from the *Visual Motion* condition and *Proprioceptive Motion* condition to obtain the *MLE* model estimates for the *Visual and Proprioceptive Motion* condition, and compared these to our empirically obtained results for the *Visual and Proprioceptive Motion* condition (see Table 10.1). A pairwise two-tailed t-test did not find significant differences between the PSEs ($t(22) = 0.7416$, $p = 0.4661$), however did find significant differences between the WFs ($t(22) = 2.7401$, $p = 0.0120$). Thus, the empirically obtained results had a larger signal dispersion than that hypothesized by the MLE model, and the MLE model did not predict how sensory integration occurred in the *Vision and Proprioceptive Motion* condition as a function of perceptual performance in each of the individual *Visual Motion* and *Proprioceptive Motion* conditions. Therefore, participants integrated the motion cues from the two sensory channels in a less efficient manner than by what the model predicted.

We propose several reasons for not observing enhanced performance in the integrated sensory condition of *Visual and Proprioceptive Motion*. First, the MLE model may not have held and sensory integration may not have been observed for our study if participants discriminated the springs using force cues rather than compliance cues. Tan et al. (1992) showed that springs can be discriminated based on force cues,

Table 10.1 WF and PSE estimates across all participants and conditions

Participant	Weber fraction [(Point of subjective equality (N/m)]			
	Visual motion	Proprioceptive motion	Visual and proprioceptive motion	Maximum-likelihood estimation model
1	0.071 (291.9)	0.040 (299.8)	0.036 (292.4)	0.035 (297.8)
2	0.152 (281.3)	0.048 (295.0)	0.061 (290.9)	0.046 (293.6)
3	0.038 (296.6)	0.056 (290.9)	0.048 (291.9)	0.031 (294.8)
4	0.059 (293.2)	0.028 (294.7)	0.047 (294.7)	0.025 (294.4)
5	0.068 (279.5)	*0.037 (291.7)	0.037 (293.0)	–
6	0.020 (290.6)	0.025 (298.7)	0.026 (296.0)	0.016 (293.7)
7	0.024 (294.1)	*0.007 (313.8)	0.007 (310.1)	–
8	0.057 (283.3)	0.031 (299.9)	0.056 (298.3)	0.027 (295.8)
9	*0.081 (288.7)	0.043 (297.4)	0.038 (300.8)	–
10	0.012 (298.2)	0.020 (296.2)	0.025 (294.8)	0.010 (297.7)
11	0.052 (296.7)	0.039 (290.0)	0.060 (297.8)	0.031 (292.3)
12	0.059 (293.7)	0.055 (285.3)	0.048 (288.5)	0.040 (289.1)
13	0.029 (290.3)	0.024 (282.2)	0.029 (283.6)	0.019 (285.4)
14	0.029 (286.8)	0.057 (300.2)	0.034 (305.9)	0.026 (289.4)
15	0.057 (287.9)	0.019 (288.8)	0.021 (295.2)	0.018 (288.7)
Mean	0.053 (290.9)	0.037 (293.5)	0.041 (294.2)	0.027 (292.7)

The mean WF and PSE values exclude Participants 5, 7, and 9, since a goodness-of-fit was not achieved for one sensory condition for each of these participants, as indicated by the asterisk (*). Adapted from Table 10.1 in Gurari et al. (2013) © IEEE 2013

Table 10.2 Largest penetration distance of a representative participant during exploration of the standard spring across all testing trials

	Largest penetration distance (cm)		
	Visual motion	Proprioceptive motion	Visual and proprioceptive motion
Minimum	0.83	0.85	0.91
Lower quartile	1.85	1.97	1.87
Median	1.99	2.09	2.05
Upper quartile	2.12	2.24	2.21
Maximum	2.48	2.61	2.96

Across all trials in each testing condition during the exploration of a standard spring for a representative participant, the respective minimum, lower quartile, median, upper quartile, and maximum distance that the finger traveled is given. Adapted from Table IV in Gurari et al. (2013) © IEEE 2013

not compliance cues, if the distance one's hand travels does not vary across testing trials, giving WFs close to 0.08. The testing procedures used and data obtained in our study do not provide conclusive evidence as to whether participants were discriminating based on force or compliance. WFs were on average less than 0.06, which correlates well with a force perception task, whereas the distance that the finger traveled was highly variable both across participants and within each participant. The largest penetration distance, or the maximum distance that the finger traveled, across all trials in each testing condition during the exploration of a standard spring for a representative participant is given in Table 10.2. The large amount of variability in the penetration distances suggests that participants were performing a compliance discrimination task. Independently of whether force or compliance discrimination was occurring, there was a need to sense the position/motion cues in order to identify the distance that the finger was traveling.

Additional possible reasons for why the MLE model did not hold, and enhanced perceptual performance was not observed in the sensory integration condition, are as follows. As the real and virtual fingers were not spatially aligned, there may be a disconnect between vision and touch. That is, the visual rendering of the user's finger as a black rotating line may map onto something other than the users finger. A further reason is that σ_V was never estimated since there was always a proprioceptive contribution; thus, it was only possible to measure σ_P and σ_{P+V} for this testing setup. Lastly, it may be that the MLE model does not describe sensory integration for this particular task.

Exploration Style

With regard to exploration style, there was no change in the interaction methods used across the different tested sensory conditions. A repeated measures one-way ANOVA was run using a Box (Greenhouse-Geisser) epsilon-hat adjustment to correct for violations of sphericity, and no significant differences were found for the largest penetration distance (F(2, 28) = 0.85, $\hat{\varepsilon}$ = 0.713, p = 0.41), number of spring presses (F(2, 28) = 0.08, $\hat{\varepsilon}$ = 0.7394, p = 0.87), and total exploration time (F(2, 28) = 1.07, $\hat{\varepsilon}$ = 0.888, p = 0.35). Thus, the exploration method employed may not necessarily change when motion cues are relayed using the different sensory pathways.

Subjective Ratings

Importantly, subjective results found that the proprioceptive motion cues were perceived as significantly more useful than the visual motion cues. Participants were asked to rank the usefulness of each of the three tested conditions. A Friedman two-way ANOVA, which is the nonparametric equivalent test to the repeated measures one-way ANOVA, was run with factors of participant and experimental sensory condition. Sensory condition significantly affected participants' usefulness ratings (χ^2(2, 28) = 16.53, p = 0.0003). A post-hoc test using a Bonferroni adjustment (MATLAB function *multcompare* which indicates significance at the p = 0.05 level) showed that the proprioceptive motion cues were perceived as significantly more useful than the visual motion cues, and the addition of vision to proprioception did not lead to perceived benefits over only proprioception. Prior to taking part in the experiment, several participants had indicated that they did not expect to find the proprioceptive motion cues to be more useful than the visual motion cues. These results suggest that the user experience of sensing a compliant object may be enhanced if motion information is relayed through the haptic channel rather than through the visual channel.

10.2.3 Summary and Implications

The motivation of this work was to identify a method for enhancing the upper-limb prosthesis experience by reducing the dependence on vision. The results are encouraging since they suggest that it may be feasible to lessen the visual burden, while maintaining comparable task performance, by shifting the reliance on vision to a reliance on proprioception for such a compliant object perception task. Findings demonstrate that compliance discrimination is possible using proprioceptive motion cues in the absence of vision. This suggests that for those experiencing compromised proprioception, it may be possible to perceive a compliant object using only haptic motion cues without taxing the visual modality. Further, given that performance did not significantly differ in the *Visual Motion* and *Proprioceptive Motion* conditions, performance should not deteriorate if visual motion cues are replaced by veridical haptic motion cues in a compliance perception task. Moreover, task performance will not necessarily improve if both cues are provided simultaneously (e.g. both vision and proprioception).

Proprioception may be subjectively valued because it is less cognitively taxing than visually observing the motion of a limb. A study by Brown et al. (2012) investigated the role of co-location of the force and position cues for a compliance perception task and showed that the coupled force/motion condition gave significantly better performance in terms of spring identification accuracy and response time than the non-coupled condition. They concluded that the uncoupled motion and force cues required the brain to remap the sensory cues to the applied motion in a manner that contradicts what may be actually occurring, perhaps either due to the disconnect between the force and motion neural pathways, or due to an integration of the stationary cues from one limb with the motion cues from the other limb. In the study presented here, for the *Proprioceptive Motion* condition the haptic cutaneous/force/torque cues combine with the haptic position/motion cues at the same location, while for the *Visual Motion* condition the haptic cutaneous/force/torque cues combine and possibly integrate across sensory modalities with the visually relayed position/motion cues. Thus, there is a physical disconnect between the cues.

Kim and Colgate (2012) investigated whether a novel multi-sensory haptic feedback device that they had developed could relay touch cues to the reinnervated chest area of upper-limb amputees who had undergone a targeted reinnervation surgical procedure [a procedure that remaps nerves that normally target the missing limb to a different site at which reinnervation occurs (Kuiken et al. 2007)]. Such an approach allowed for the artificially relayed sensory sensations to match both somatotopically and by modality. That is, respectively the location and the sensation that was artificially conveyed matched the location and the sensation that was being stimulated on the prosthetic limb. Results demonstrate that haptically displaying information about a single modality (shear or pressure feedback) successfully decreased forces applied when grasping an object, while the combination of relaying both sensations gave higher grasping forces than when relying on only one of the sensations. These findings demonstrate that there may be a limit to the amount of information that can be effectively transmitted, even when using somatotopic and modality matching.

In the next section, we provide an overview of methods by which compliance could be relayed artificially using haptic artificial proprioception displays. Implications from our study, (Brown et al. 2012), and (Kim and Colgate 2012) suggest that artificial cues should be relayed in a manner that is haptically natural and intuitive in order to maximize user performance and ease of use.

10.3 Haptic Sensory Substitution Systems for Compliance Perception

Here we provide an overview of tactile feedback technologies and considerations for relaying compliance using sensory substitution systems. Sensory substitution is defined as the mapping of one sensation (e.g. proprioception) to another (e.g. vibration) in order to convey the original information in a different way. One example of a very effective and well-known device that functions as such is the cell phone, which uses vibrations to notify the user of an incoming call. It is a small and light-weight packaged system, and it effectively uses vibratory cues to convey what was previously auditory stimulation. Below we give an overview of haptic feedback sensory substitution devices, discuss practical locations for relaying the sensations, and provide an overview of work already done in relaying compliance, force, and position information.

10.3.1 Challenges in Sensory Substitution

Given that the mechanism by which compliance perception occurs is not completely known, it is not clear how this sensation should be artificially relayed. Once a model that explains how compliance perception occurs is validated, one can go about the task of relaying the missing haptic information using artificial feedback devices. For example, tactile cues were demonstrated to be sufficient for sensing springs with deformable surfaces (Srinivasan and LaMotte 1995). In turn, a novel device was created to interact with the finger cutaneously by increasing a contact surface area as a function of an applied force. The combination of the tactile feedback with the kinesthetic feedback gave, as expected, better compliance perception than relying solely on kinesthetic cues (Bicchi et al. 2000). The research direction on which we focus is the interaction with rigid-surfaced springs, where both kinesthetic and tactile cues are important for compliance perception (Srinivasan and LaMotte 1995).

We simplify the task to one of interpreting the kinesthetic cues [force and position (Chib et al. 2009)], ignoring the tactile cues, to demonstrate the complexity of this work. Some open-ended questions for relaying kinesthetic compliance cues include: should both force and position information be relayed, or only one of these? How will the increased cognitive load from using the sensory substitution system impact task performance? When and how should the sensory information be delivered? For example, should position and/or velocity be conveyed? Finally, should the device be activated at all times, or should it be only activated to indicate changes between an initial position and a final target position when exploring an object?

We demonstrate the difficulty of artificially relaying compliance by discussing the example of conveying the compliance sensed by one hand. The human hand has 22+ degrees-of-freedom (DoF). Compliance information could be displayed by artificially stimulating the user to relay a net force using a device such as a linear actuator and a separate device to portray the net hand position. However, by reducing the information displayed, the compliance sensed at each unique finger is not necessarily being represented. If one attempts to relay all of the information, an immense number of devices would be required to represent each DoF and sensory cue (e.g. force, position).

An additional challenge is artificially displaying the information in a manner that will not interfere with one's ability to perform activities of daily living. That is, sensory substitution mechanisms should be usable during the wearer's daily routine.

For example, users should be able to sense a change in position and force, integrate the information to sense an object's compliance, and simultaneously actively maintain currently relevant information in the mind (e.g. working memory load).

Using tactile sensory substitution systems is beneficial compared to relying on visual and auditory feedback since areas on the body that are not generally used during activities of daily living can be stimulated. Such systems are comprised of one or more tactors—elements that stimulate the user via the tactile channel—to convey information. Tactors are challenging to implement because it is necessary to identify an effective stimulation location on the body; select a design and control schema for stimulating the user; discern which information is critical to feed back to the user; package the system into a compact, light-weight, portable system that responds in real time; select an appropriate stimulation method (e.g. vibrotactile, electrocutaneous cues); and ensure that the artificial stimulation will not interfere with one's daily activities (Jones and Sarter 2008; MacLean 2009; Stanley and Kuchenbecker 2011).

10.3.2 Haptic Stimulation

Artificial haptic stimulation has been implemented both through invasive and non-invasive means. Ideally, these sensory feedback systems would have somatotopic matching, where the bodily location is directly stimulated such that the feeling is natural and intuitive; however, this may require an invasive procedure such as targeted reinnervation (Kim and Colgate 2012). An additional item for consideration is matching the sensory modality, such that the sensation relayed to the user is identical to that which was externally perceived, so that feeling may be more easily and quickly interpreted (Kim and Colgate 2012).

Tested invasive stimulation methods include using targeted reinnervation (Kuiken et al. 2007) and peripheral nerve stimulation (Dhillon and Horch 2005; Horch et al. 2011; Riso 1999). Such invasive stimulation methods are not yet developed technologically to safely, naturally, and seamlessly relay realistic compliance, force, and position cues.

A more feasible method for actively displaying haptic information currently is the use of noninvasive methods such as electrical, vibratory, force, and skin stretch stimulation. An article by Kaczmarek et al. (1991) provides a thorough overview of the first two sensory stimulation methods. They also provide a list of performance criteria considerations for such devices, which includes minimal power consumption, maximal stimulation comfort, minimal post stimulation skin irritation, minimal sensory adaption, and maximal information transfer. Note that training may be necessary to remap the neural pathways for interpreting the new flow of sensory information.

Below we give an overview of numerous novel haptic feedback technologies that have been developed in recent years, and in Fig. 10.4 we depict several sensory substitution devices that we have tested for relaying compliance and proprioception artificially to various locations on the body. Methods tested for artificially displaying compliance-related information include using multiple vibrating elements for con-

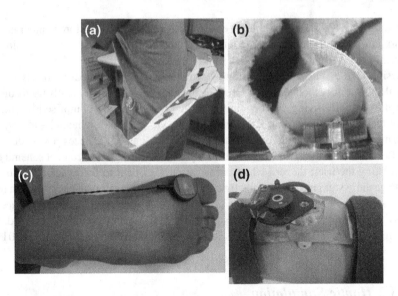

Fig. 10.4 Examples of sensory substitution devices. **a** Vibratory feedback via multiple vibrating elements to the torso (adapted from Fig. 10.3 in Cheng (2012) © IEEE 2012). **b** Skin slip feedback via a single rotating element with a raised ridge. **c** Vibratory feedback to the toe (adapted from Fig. 10.3 in Gurari et al. (2009) © IEEE 2009), and **d** skin stretch feedback to the forearm (adapted from Fig. 10.2 in Gurari et al. (2012) © Springer 2012)

veying hand configuration to the torso (Cheng 2012), a novel skin stretching/slipping mechanism for relaying angular position to the big toe, a comparison of perception of vibration at the finger, foot, and forearm (Gurari et al. 2009), and stretching the skin on the user's forearm to display proprioception artificially during compliance perception (Gurari et al. 2012).

10.3.2.1 Feedback Mechanisms

Currently, the most common method for conveying information haptically is using vibration feedback. Vibratory feedback devices are cheap, have relatively low power consumption, convey highly perceivable signals, and are small, light-weight, and portable. Drawbacks include the limited information that can be displayed using one vibrating element, possible confusion of interpreting multiple vibrating elements, discomfort and/or annoyance of the sensation, and disregard of the signal if it is continuous. Pylatiuk et al. (2006) suggested that vibratory feedback be limited to between 3 and 5 s of continuous stimulation in order to limit user annoyance. A detailed discussion of the advantages and disadvantages of vibrotactile feedback devices is given in Stanley and Kuchenbecker (2011).

Another widely tested method for relaying information haptically is electrical stimulation (Cholewiak and Collins 2003; Kajimoto et al. 1999; Riso and Ignagni 1985; Rohland 1975). Benefits of using electrocutaneous signals are similar to those of vibrotactile stimulation, and include low cost, light weight, compact size, and low power consumption. A major drawback of this stimulation modality is that the cues can be uncomfortable, if not painful, especially if the stimulated skin is moist.

Force feedback has also been used for conveying sensory cues artificially (Kim et al. 2010; Panarese et al. 2009; Prattichizzo et al. 2012). Force, or pressure, feedback can be beneficial because the mapping can be more intuitive if the force being measured at one location is artificially displayed through a force-feedback mechanism at another location. Additional benefits of force feedback include that it can be a relatively lightweight and compact device (although larger and heavier than vibratory and electrocutaneous devices), requires low power consumption, and is low in cost. A drawback is that the amount of information that can be relayed through a single device is limited. Note that force feedback can be considered as complimentary to vibrotactile feedback, since the difference between the two sources of information is in the frequency, the former exhibiting a lower frequency response and the latter exhibiting a higher frequency response.

Skin stretch is a more recently explored idea that is being pursued for conveying sensory information, e.g. (Bark et al. 2008; Gleeson et al. 2009). Skin stretch feedback is useful because it can give position, velocity, and direction information all from one mechanism. Additionally, this method has been shown to give better performance in displaying position and velocity cues over a vibrotactile stimulus (Bark et al. 2008). A limitation of this method is that the information will not be effectively conveyed if the end-effector to skin contact is not strong, such as when the skin is moist and/or the adhesive material used is not strong enough.

Electrocutaneous, vibrotactile, force, and skin stretch stimulation are stimulation methods that have all been applied for artificially conveying compliance, force, and/or position information (Bark et al. 2008; Cipriani et al. 2008; Kim et al. 2010; Mann and Reimers 1970; Panarese et al. 2009; Pylatiuk et al. 2006; Riso and Ignagni 1985; Rohland 1975; Schorr et al. 2013; Wheeler et al. 2009). Mixing numerous feedback signals into one device may enable richer information to be simultaneously conveyed, as suggested by Kim et al. (2010) when they created a novel multi-sensory feedback device that displays touch, pressure, vibration, shear force, and temperature all-in-one. Another idea proposed for realistically displaying sensations is to have body-machine interfaces (Casadio et al. 2010). Such interfaces allow a patient to use body parts that are still partially functional in order to control machines that assist them in performing tasks. These are valuable, in part, because they allow for natural proprioceptive cues at another portion of the body to be remapped to proprioceptive cues at the missing sensory channel.

10.3.2.2 Feedback Locations

The sensitivity of different locations on the body to various haptic stimuli depends on which location of the body is being stimulated. The glabrous (hairless) skin at locations such as one's toes, fingertips, or lips has a greater sensory bandwidth, and in turn, is more sensitive than hairy skin regions at locations such as one's back, shoulder, and forearm (Weinstein 1968). Sensitivities vary depending on the available mechanoreceptors, the response properties of their afferents (Merzenich and Harrington 1969), and their corresponding density levels. The more sensitive glabrous skin areas tend to be employed for activities of daily living, making these sites less ideal for use in sensory substitution systems.

We conducted several studies to identify the effectiveness of various sensory substitution devices for relaying information to different parts of the body, focusing primarily on two locations: the foot and the torso (see Fig. 10.4). The foot is of interest because it has the highly sensitive glabrous skin and, thus, offers a relatively large surface area of mechanoreceptors. Additionally, it provides the possibly convenient option of developing a system that can be packaged inside a shoe. The torso is of interest since it is a portion of the body that is not actively used during activities of daily living, and it has a large surface area that could allow for the placement of a large number of actuators. Furthermore, packaging for the system can be placed underneath a shirt so that it is hidden from view.

The effectiveness of various locations for haptic sensory substitution systems on the body has been tested in many scenarios. For example, prior research demonstrated that the vibrotactile stimulation of the torso is easier to perceive than stimulation of the forearm (Jones et al. 2006), and another study demonstrated that perception of vibratory cues was comparable at the glabrous sites of the fingertip and foot, which were both superior in performance to the forearm (Gurari et al. 2009).

10.3.2.3 Relaying Compliance Through Haptic Sensory Substitution Systems

Vibrotactile, electrical and force stimulation have all been used for displaying force and position cues artificially. Several decades ago, work done by Shannon (1976) and Scott et al. (1980) demonstrated that vibrotactile stimulation could be used to convey grip/pinch force to upper-limb prosthesis users, while (Mann and Reimers 1970) used this stimulation modality for displaying the angular position of the elbow joint of a prosthesis. More recently, (Pylatiuk et al. 2006) and (Chatterjee 2007) showed that vibratory stimulation can aid in controlling the grasping force of a myoelectrically controlled prosthetic device, and (Cipriani et al. 2008) found that vibrotactile feedback was subjectively important to users, even though it did not improve performance during an object grasping task. Additionally, force feedback to the toes has been effective for relaying grip force sensed by a robotic hand (Panarese et al. 2009). Furthermore, electrical stimulation was used to convey the position of a user in a motion-tracking task, and it was shown to give improved task performance over no feedback, but worse performance than when relying on visual cues (Riso and Ignagni 1985). Another study used electrical stimulation for relaying gripping pressure when grasping an object, and they concluded that the electrical stimulation is desired by upper-limb prosthesis users for such a task (Rohland 1975).

Very recent work investigated whether compliance could be relayed with vibrotactile feedback of force and position information (Witteveen et al. 2013). A single high-quality vibrating element modulated amplitude to convey either force or position information while the latter modulated the activated position of 4–8 cheaper vibrating elements to relay the second piece of compliance information. Results demonstrate that compliance perception was most effective when both force and proprioceptive cues were artificially portrayed and when solely artificial proprioceptive information was conveyed. The artificial display of solely force cues resulted in significantly worse task performance. Interestingly, response times for interpreting compliance were shorter with artificially conveyed proprioceptive cues over the combination. This is similar to the finding in Kim and Colgate (2012) in that the combination of pressure and shear feedback gave worse task performance over receiving information from only one of the sensory channels.

Currently, skin stretch stimulation methods are receiving attention for the display of position, force, and compliance. Using such a display may be more intuitive than the aforementioned methods for several reasons. Firstly, skin stretch is a natural signal used in sensing proprioception and force. Secondly, there may be more natural mappings (such as rotation or translation of a device for a rotary limb motion or linearly applied force), and finally, a large amount of information may be more easily conveyed using a single device (e.g. position, motion, and directional information all-in-one). Schorr et al. (2013) found that artificially relayed skin stretch cues can give comparable compliance discrimination performance to force feedback cues, and the mapping between the skin stretching and the signal is intuitive (no training is required). Using a different skin stretching device, (Bark et al. 2008) demonstrated that skin stretch feedback was more effective at conveying proprioception during a sighted targeting task over vibratory feedback, which was superior to having no artificial haptic feedback cues.

Below we describe two relevant studies conducted by the authors: one identified the effectiveness of a skin stretching mechanism (Wheeler et al. 2009) for artificially displaying compliance (Sect. 10.3.3), and the second investigated the feasibility of conveying hand configuration by relaying four synergistic motions of the hand via four vibrating elements (Sect. 10.3.4).

10.3.3 Experiment: Relaying Compliance Using Skin Stretch Feedback

In Sect. 10.2.2 we presented an experiment demonstrating that the perception of a virtual compliant object is similar in the cases when motion cues are perceived visually or proprioceptively. Here, we investigate perceptual performance in the compliance discrimination task when motion cues are also conveyed using an artificial proprioceptive display.

10.3.3.1 Experimental Overview

We used the skin stretch device presented in Wheeler et al. (2009) and shown in Fig. 10.4d to repeat the study described in Sect. 10.2.2 under slightly modified experimental procedures. Visual feedback was relayed on the computer monitor, but here the monitor was horizontally aligned with the real finger (via visual inspection) and vertically aligned such that the virtual finger was offset by approximately 0.09 m from the real finger. Below we provide a brief overview of this work; the experimental setup and study is explained in more detail in Gurari et al. (2012).

The skin stretch device was designed to stretch the user's skin clockwise and counterclockwise to a maximum range-of-motion of 40° in each direction. The device attaches to the user's right forearm via velcro straps, and two cylindrical 1.4 cm disks that are spaced 2.6 cm apart make contact with the user's skin. The device was comprised of a Shinesei USR30-B3 non-backdriveable ultrasonic motor (Himeji, Japan) that applies a torque to the end-effector via a 6:1 capstan/cable transmission, and an optical encoder that measures the angular rotation of the end-effector. Shape Deposition Manufacturing (Binnard and Cutkosky 2000) was used to manufacture the body of the device.

Eight healthy, intact participants were involved in the study. They used the custom kinesthetic feedback haptic device shown in Fig. 10.1 to control the motion of a real and/or virtual limb by applying a force at the right index finger. Thus, an effort was always actively applied, producing corollary discharge cues, and forces applied at the index finger were always sensed. Participants performed a method of constant stimuli, two-alternative forced choice task on rigid-surfaced virtual springs. They compared springs with a reference stiffness of 290 N/m to one of 10 comparison springs with a stiffness ranging between 250 N/m and 330 N/m in increments of 8 N/m. Motion feedback was portrayed under four possible conditions: vision (*Vision*), natural proprioception (*Proprioception*), artificial proprioception (*Skin Stretch*), or artificial proprioception combined with sight (*Skin Stretch with Vision*).

The aim of this study was to quantify perceptual performance across each of the conditions to determine the effectiveness of the artificial proprioception device in comparison with vision and natural proprioception. The artificial proprioceptive condition used the skin stretching device to stretch the skin on one's ipsilateral forearm by an amount proportional to the angular rotation of a virtual prosthesis. In Fig. 10.5, we pictorially depict the tested conditions and indicate how a user of a myoelectric upper-limb prosthesis, and a user of this experimental setup, may combine and integrate sensory cues to perceive the artificial limb's location.

Participants interacted with the springs for a maximum of 30 s [which was demonstrated to be a reasonable amount of time for comfortable exploration (Gurari et al.

Experimental Sensory Conditions

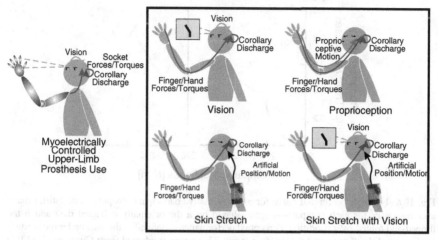

Fig. 10.5 Sensory cues available during the experimental testing conditions and as compared to myoelectrically controlled upper-limb prosthesis use. Adapted from Fig. 10.1 in Gurari et al. (2012) © Springer 2012

2009)] and were asked to indicate which spring was stiffer. The maximum force was obtained during a pre-testing calibration procedure based on the maximum amount of force that the user could comfortably apply. This was to ensure that the entire skin stretch device workspace was effectively employed. Additionally, the interaction speed was limited to 100°/sec to ensure that spring stiffness rendered using the kinesthetic feedback device was comparable to those rendered visually and artificially (Gurari et al. 2013).

As described in Sect. 10.2.2.1, the different sensory conditions result in different afferent pathways of the motion cues. The manner by which the corollary discharges combine with the force cues remains comparable across conditions, whereas the motion is perceived as follows for each of the conditions. During *Proprioception*, the user's finger rotates by an amount that is proportional to his/her applied force. During *Vision*, a graphical representation of the virtual finger, as depicted by a vertical line that is the length of the user's real finger, rotates by an amount that is proportional to the user's applied force (varies as a function of the rendered spring) while the real finger is held still. During *Skin Stretch*, the skin stretch device rotates on the user's forearm by an amount that is proportional to his/her applied force (varies as a function of the rendered spring) to indicate the location of the virtual finger while the real finger is held still. During *Skin Stretch with Vision*, the user observes a graphical representation of the virtual finger rotating on the monitor and feels the skin stretch device rotating on his/her forearm by an amount that is proportional to the applied finger force (varies as a function of the rendered spring) while the real finger is held still. For all conditions, aside from

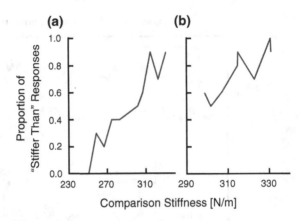

Fig. 10.6 Example perceptual curve for a representative participant. Proportion of "stiffer than" user responses for each comparison spring based on **a** the originally collected data and **b** the flipped-and folded-over condition. Perceptual performance is defined as the area under the normalized curve of the flipped-and folded-over curve. This figure is adapted from Gurari et al. (2012) © Springer 2012

Proprioception, sensory integration may occur between the visually and/or haptically relayed cues and the stationary positional cues from the real finger.

10.3.3.2 Results and Discussion

The effect of sensory condition on task performance and exploration methods was evaluated using a repeated measures one-way ANOVA with a Box (Geisser-Greenhouse) epsilon-hat adjustment to correct for violations of sphericity. Perceptual performance is defined here as the area under the normalised curve, since it describes the cumulative discrimination capabilities and allows for a comparison across conditions (see Fig. 10.6). The WF was not used since a goodness-of-fit was not obtained for a number of the conditions and participants given the employed analysis method. Sensory condition was not found to significantly affect perceptual performance ($F(3, 21) = 0.84$, $\hat{\varepsilon} = 0.5134, p = 0.43$) and exploration method [number of spring presses ($F(3, 21) = 0.42$, $\hat{\varepsilon} = 0.5800, p = 0.64$), total time spent pressing springs ($F(3, 21) = 1.54$, $\hat{\varepsilon} = 0.6964, p = 0.25$), or largest penetration distance ($F(3, 21) = 1.18$, $\hat{\varepsilon} = 0.5191, p = 0.33$)].

Participants provided feedback on how useful they perceived each feedback condition, as well as how difficult they found the stiffness perception task using each sensory feedback condition. The Friedman test (factors of participant and sensory condition) did not find an effect of sensory condition on perceived usefulness rankings ($\chi^2(3) = 4.95, p = 0.176$); however, perceived task difficulty was significant ($\chi^2(3) = 9.64, p = 0.022$) and a post-hoc analysis using the Tukey's honestly significant difference criterion found a significant difference between proprioceptive cues only and skin stretch cues only at a significance level of $p = 0.05$ (using the MATLAB function *multcompare*). A possible drawback of the artificial skin stretch cues is that the task may be perceived as more difficult when it is used. The force cues combined with the position cues conveyed from the skin stretching mechanism were sufficient for discriminating between springs; however, the increase in task difficulty may be practically problematic.

A general perception is that haptic cues are required in order to control one's motions, e.g. (Cole 1995). Interestingly, these results along with the findings from the previously presented study demonstrate that vision and proprioception do not give significantly different performance results. Future work could better inform on the role of the natural and artificial cues by identifying task performance and subjective feedback for a passive task when compared to this active task. Such testing may additionally inform on the effectiveness of the artificial cues for scenarios when an artificial limb may automatically perform a task (e.g. automatic grasping of an object). Further, studies could test the system with amputees to identify its potential as well as limitations for artificially relaying proprioception (and/or other haptic cues) in practice, since participants in this study were all intact healthy individuals. Moreover, testing with additional participants on the current setup could give higher confidence in the findings.

10.3.4 Experiment: Perception of Hand Proprioception via Multiple Vibrating Elements

10.3.4.1 Experimental Overview

It is challenging to create a sensory substitution haptic display for relaying compliance for practical purposes. Ideally, either force or position cues will be an "input", and only one of these sensations will be artificially displayed to convey the "output" of the compliant object. It is not clear, however, how to create a device that can display sensory information in a manner that is comparable to natural proprioception in terms of information transfer and cognitive loading, especially if the stimulation is noninvasive. Furthermore, being able to display information for higher-DoF tasks adds significantly more mechanical and control complexity. A clever and intuitive mapping should be developed that can effectively relay the position and/or force cues perceived at all relevant digits. Proposed possible algorithms for how this could be done include conveying the entire hand posture based on measurements of solely the fingertip positions (Mulatto et al. 2010) or fusing a priori information with a limited and possibly noisy set of sensor data (Bianchi 2012).

The authors recently investigated the possibility of conveying a reduced set of proprioceptive information about the 22-DoF hand using a subset of four unique hand configurations, or hand synergies (Cheng 2012). This work is motivated by the idea that nearly 85 % of one's hand motions can be broken down into four unique hand movements (Thakur et al. 2008). Thus, by displaying a subset of sensory information, it may be possible to sense and, in turn, control one's hand movements.

In the hand-synergy feedback experiment, four vibrotactile devices around the torso were activated in controlled vibration patterns to convey hand configuration (see Fig. 10.4b). The vibrotactile elements were activated via two unique mappings to identify each one's effectiveness: (1) a Synergy-based mapping motivated by results of a principal component analysis on general hand motions, and (2) Decoupled hand motions that were experimenter-chosen and included finger grasp and finger spread. Each tactor could be activated at one of three levels in which the envelope frequency was respectively either 0.5, 1.0, or 2.0 Hz at a carrier frequency of 250 Hz.

Fourteen participants took part in the experiment, all healthy with no neurological illnesses. Six participants were female, eight male, and all provided informed consent. The task was to identify one of five possible hand configurations using either the Synergy or the Decoupled tactor stimulation patterns. The experiment spanned two days. On the first day, half of the participants interacted with the Synergy method and the other half on the Decoupled method. On the second day, the stimulation method applied was switched. Participants trained on the system with visual feedback for 30 min prior to the testing to learn the mapping between the tactile stimulation and the hand configuration. Data collection for the testing portion was conducted when only vibrotactile cues were relayed to convey the hand configuration.

10.3.4.2 Results and Discussion

Results demonstrate that there is still a long way to go for interpreting the best mapping from a multiple sensory space to effectively relaying the information using artificial haptic display devices. Tactor patterns were correctly identified 74.3 % of the time for the Synergy method and 53.6 % of the time for the Decoupled method. The ability to correctly identify the tactor pattern along with the corresponding hand configuration was 68.6 % for the Synergy method and 46.4 % for the Decoupled method, where guessing, or chance, was 2.5 %. User response times were much less inspiring, with participants requiring 30–45 s to estimate the hand configurations. Comparisons were made for the two methods—Synergy and Decoupled—using a repeated measures one-way ANOVA. The ability to correctly identify hand configurations was significantly higher ($F(1,12) = 16.91$, $p = 0.001$), time elapsed to identify the hand configurations was significantly less ($F(1, 12) = 26.67, p = 0.0002$), and ability to identify solely the tactor pattern was significantly better ($F(1, 12) = 6.72, p = 0.023$) for the Synergy method than the Decoupled method. Future work for relaying sensory information should ensure that both the tactor patterns and the sensory cues are as orthogonal, or as different, as possible to enhance user interpretation of the signals.

10.4 Concluding Remarks

In the first part of the chapter, we underscored the importance of proprioception for compliance perception. An experiment was presented that demonstrated that the combination of force and proprioception cues allows for compliance perception, and that sensing the force and position information haptically is less cognitively taxing than the combination of a haptically relayed force and a visually observed motion of a limb.

In the second part of the chapter, we proposed touch feedback devices that could be used for artificial display of proprioception to those with compromised touch sensing (e.g. lack of proprioception). Skin stretch is a promising method because it may allow for more natural mappings (e.g. rotation of a device to portray a rotary limb motion) and the information displayed may better encompass the pertinent information (e.g. ability to display position, motion, and directional cues simultaneously with one device). Experimental results indicate that task performance may be possible using skin stretch feedback. There are still limitations, however, since the task may be perceived to be more difficult using such cues.

For future work, an enhanced understanding of the neural mechanisms underlying compliance perception may provide inspiration for how to artificially relay missing information intuitively and naturally. If the mechanism governing how human position, force, and tactile cues merge to create the compliance percept is unlocked, then perhaps simpler artificial stimulation patterns can be generated to involve the appropriate neural pathways to create the desired sensations. Furthermore, this enhanced understanding could be eventually applied to patient populations with limited or nonexistent sensing to aid in their ability to move around and interact with their surroundings.

Acknowledgments The authors would like to acknowledge Katherine J. Kuchenbecker, Xin Alice Wu, Caroline A. Montojo, Jason Wheeler, Amy Shelton, Andrew Cheng, Kirk A. Nichols, Heidi M. Weeks, and Steven Hsiao for their contributions to the discussion and experiments presented in this chapter. Additionally, the authors thank Alessandra Sciutti for reviewing and providing feedback on the manuscript. Financial support for the work presented here was provided by a National Science Foundation Graduate Fellowship, the Johns Hopkins University Applied Physics Laboratory under the DARPA Revolutionizing Prosthetics program, contract N66001-06-C-8005, Johns Hopkins University Brain Science Institute, a travel award from the IEEE Technical Committee on Haptics, Stanford University, and Istituto Italiano di Tecnologia.

References

Atkins DJ, Heard DCY, Donovan WH (1996) Epidemiologic overview of individuals with upper-limb loss and their reported research priorities. J Prosthet Orthot 8(1):2–11

Avanzini F, Crosato P (2006) Haptic-auditory rendering and perception of contact stiffness. In: Proceedings of the first international conference on haptic and audio interaction design (HAID'06), pp 24–35

Bark K, Wheeler JW, Premakumar S, Cutkosky MR (2008) Comparison of skin stretch and vibrotactile stimulation for feedback of proprioceptive information. In Proceedings of the 16th international symposium on haptic interfaces for virtual environments and teleoperator systems, pp 71–78

Bastian HC (1887) The muscular sense; its nature and cortical localisation. Brain 10:1–89

Berryman LJ, Yao JM, Hsiao SS (2006) Representation of object size in the somatosensory system. J Neurophysiol 96(1):27–39

Bianchi M, Salaris P, Bicchi A (2012) Synergy-based optimal design of hand pose sensing. In 2012 IEEE/RSJ international conference on intelligent robots and systems (IROS), pp 3929–3935

Bicchi A, Scilingo EP, Rossi DD (2000) Haptic discrimination of softness in teleoperation: the role of the contact area spread rate. IEEE Trans Robot Autom 16(5):496–504

Binnard M, Cutkosky MR (2000) Design by composition for layered manufacturing. J Mech Des 122(1):91–101

Blank A, Okamura AM, Kuchenbecker KJ (2010) Identifying the role of proprioception in upper-limb prosthesis control: studies on targeted motion. ACM Trans Appl Percept, 7(3):15.1–15.23

Brewer BR, Fagan M, Klatzky RL, Matsuoka Y (2005) Perceptual limits for a robotic rehabilitation environment using visual feedback distortion. IEEE Trans Neural Syst Rehabili Eng 13(1):1–11

Brown JD, Gillespie RB, Gardner D, Gansallo, EA (2012) Co-location of force and action improves identification of force-displacement features. In: Proceedings of the IEEE haptics symposium, pp 187–193

Burke RE (2007) Sir charles sherrington's the integrative action of the nervous system: a centenary appreciation. Brain 130:887–894

Casadio M, Pressman A, Fishbach A, Danziger Z, Acosta S, Chen D et al (2010) Body machine interface: remapping motor skills after spinal cord injury. Exp Brain Res 207:233–247

Chatterjee A (2007). Vibrotactile haptic feedback for advanced prostheses. Unpublished master's thesis, Department of Biomedical Engineering, Johns Hopkins University, 3400 N. Charles Street, Baltimore, MD 21218

Cheng A, Nichols KA, Weeks HM, Gurari N, Okamura AM (2012) Conveying the configuration of a virtual human hand using vibrotactile feedback. IEEE Haptics Symposium, pp 155–162

Chib VS, Krutky MA, Lynch KM, Mussa-Ivaldi FA (2009) The separate neural control of hand movements and contact forces. J Neurosci 29(12):3939–3947

Cholewiak RW, Collins AA (2003) Vibrotactile localization on the arm: Effects of place, space, and age. Percept Psychophys 65(7):1058–1077

Cipriani C, Zaccone F, Micera S, Carrozza MC (2008) On the shared control of an EMG-controlled prosthetic hand: Analysis of user-prosthesis interaction. IEEE Trans Robot 24(1):170–184

Cole J (1995) Pride and a daily marathon. The MIT Press, Massachusetts

Collins DF, Prochazka A (1996) Movement illusions evoked by ensemble cutaneous input from the dorsum of the human hand. J Physiol 496(3):857–871

Collins DF, Refshauge KM, Todd G, Gandevia SC (2005) Cutaneous receptors contribute to kinesthesia at the index finger, elbow, and knee. J Neurophysiol 94(3):1699–1706

Crapse TB, Sommer MA (2008) Corollary discharge across the animal kingdom. Nat Rev 9(8): 587–600

Dhillon GS, Horch KW (2005) Direct neural sensory feedback and control of a prosthetic arm. IEEE Trans Neural Syst Rehabilit Eng 13(4):468–472

Edin BB (2004) Quantitative analyses of dynamic strain sensitivity in human skin mechanoreceptors. J Neurophysiol 92(6):3233–3243

Ernst MO, Banks MS (2002) Humans integrate visual and haptic information in a statistically optimal fashion. Nature 415(6870):429–433

Gandevia SC, Smith JL, Crawford M, Proske U, Taylor JL (2006) Motor commands contribute to human position sense. J Physiol 571(3):703–710

Gescheider GA (1997) Psychophysics: the fundamentals, 3rd edn. Lawrence Erlbaum Associates Inc, UK

Ghez C, Gordon J, Ghilardi MF, Christakos CN, Cooper SE (1990) Roles of proprioceptive input in the programming of arm trajectories. In Cold spring harbor symposia on quantitative biology, pp 837–847

Gleeson BT, Horschel SK, Provancher WR (2009) Communication of direction through lateral skin stretch at the fingertip. Third joint eurohaptics conference and symposium on haptic interfaces for virtual environment and teleoperator systems, pp 172–177

Gurari N, Kuchenbecker KJ, Okamura AM (2009) Stiffness discrimination with visual and proprioceptive cues. In Proceedings of the third joint eurohaptics conference and symposium on haptic interfaces for virtual environment and teleoperator systems (ieee world haptics), pp 121–126

Gurari N, Kuchenbecker KJ, Okamura AM (2013) Perception of springs with visual and proprioceptive motion cues: Implications for prosthetics. IEEE Trans Hum Mach Syst 43(1):102–114

Gurari N, Smith K, Madhav M, Okamura AM (2009) Environment discrimination with vibration feedback to the foot, arm, and fingertip. In 11th international conference on rehabilitation robotics, pp 343–348

Gurari N, Wheeler J, Shelton A, Okamura AM (2012) Discrimination of springs with vision, proprioception, and artificial skin stretch cues. In Isokoski P, Springare J (Eds.), Haptics: perception, devices, mobility, and communication (Vol. 7282, pp. 160–172). Springer, Berlin Heidelberg

Horch K, Meek S, Taylor TG, Hutchinson DT (2011) Object discrimination with an artificial hand using electrical stimulation of peripheral tactile and proprioceptive pathways with intrafascicular electrodes. IEEE Trans Neural Syst Rehabilit Eng 19(5):483–489

Jones LA, Lockyer B, Piateski E (2006) Tactile display and vibrotactile pattern recognition on the torso. Adv Robot 20(12):1359–1374

Jones LA, Sarter NB (2008) Tactile displays: guidance for their design and application. Hum Factors 50:90–111

Kaczmarek KA, Webster JG, Bach-y-Rita P, Tompkins WJ (1991) Electrotactile and vibrotactile displays for sensory substitution systems. IEEE Trans Biomed Eng 38(1):1–16

Kajimoto H, Kawakami N, Maeda T, Tachi S (1999) Tactile feeling display using functional electrical stimulation. In: Proceedings of the 9th international conference on artificial reality and telexistence (ICAT), pp 107–114

Kelso JAS (1977) Motor control mechanisms underlying human movement reproduction. J Exp Psychol Hum Percept Perform 3(4):529–543

Kim K, Colgate JE (2012) Haptic feedback enhances grip force control of semg-controlled prosthetic hands in targeted reinnervation amputees. IEEE Trans Neural Syst Rehabil Eng 20(6):798–805

Kim K, Colgate JE, Santo-Munné JJ, Makhlin A, Peshkin MA (2010) On the design of miniature haptic devices for upper extremity prosthetics. IEEE/ASME Trans Mechatron 15(1):27–39

Kuiken TA, Marasco PD, Lock BA, Harden RN, Dewald JPA (2007) Redirection of cutaneous sensation from the hand to the chest skin of human amputees with targeted reinnervation. Proc the Nat Acad sci United States of America 104(50):20061–20066

Kuschel M, Luca MD, Buss M, Klatzky RL (2010) Combination and integration in the perception of visual-haptic compliance information. IEEE Trans Haptics 3(4):234–244

Larish DD, Volp CM, Wallace SA (1984) An empirical note on attaining a spatial target after distorting the initial conditions of movement via muscle vibration. J Mot Behav 16(1):76–83

Lecuyer A, Coquillart S, Kheddar A, Richard P, Coiffet P (2000) Pseudohaptic feedback: Can isometric input devices simulate force feedback? In: Proceedings of the ieee virtual reality, pp 83–90

MacLean KE (2009) Putting haptics into the ambience. Trans Haptics 2(3):123–135

Mann RW, Reimers SD (1970) Kinesthetic sensing for the EMG controlled Boston Arm. IEEE Trans Man Mach Syst 11:110–115

Matthews PBC (1982) Where does sherrington's muscular sense originate? muscles, joints, corollary discharges. Ann Rev Neurosci 5:189–218

McCloskey DI (1978) Kinesthetic sensibility. Physiolo Rev 58(4):763–820

Merzenich MM, Harrington T (1969) The sense of flutter-vibration evoked by stimulation of the hairy skin on primates: comparison of human sensory capacity with the responses of mechanoreceptive afferents innervating the hairy skin on monkeys. Exp Brain Res 9:236–260

Mulatto S, Formaglio A, Malvezzi M, Prattichizzo D (2010) Animating a synergy-based deformable hand avatar for haptic grasping. Haptics: generating and perceiving tangible sensations lecture notes in computer science, vol 6192. Springer, Berlin Heidelberg, pp 203–210

Paljic A, Burkhardt J-M, Coquillart S (2004) Evaluation of pseudo-haptic feedback for simulating torque: a comparison between isometric and elastic input devices. In: Proceedings of the 12th international symposium on haptic interfaces for virtual environment and teleoperator systems, pp 216–223

Panarese A, Edin BB, Vecchi F, Johansson RS (2009) Humans can integrate force feedback to toes in their sensorimotor control of a robotic hand. IEEE Trans Neural Syst Rehabil Eng 17(6):560–567

Prattichizzo D, Pacchierotti C, Rosati G (2012) Cutaneous force feedback as a sensory substraction technique in haptics. IEEE Trans Haptics 5(4):289–300

Pressman A, Welty LJ, Karniel A, Mussa-Ivaldi FA (2007) Perception of delayed stiffness. Int J Robot Res 26(11–12):1191–1203

Pylatiuk C, Kargov A, Schulz S (2006) Design and evaluation of a low-cost force feedback system for myoelectric prosthetic hands. J Prosthet Orthot 18(2):57–61

Ramachandran VS, Rogers-Ramachandran D (1996) Synaesthesia in phantom limbs induced with mirrors. Pro Royal Soc London Ser B Biolo Sci 263(1369):377–386

Ramachandran VS, Rogers-Ramachandran D, Cobb S (1995) Touching the phantom limb. Nature 377:489–490

Riso RR (1999) Strategies for providing upper extremity amputees with tactile and hand position feedback - moving closer to the bionic arm. Technol Health Care 7(6):401–409

Riso R R, Ignagni AR (1985) Electrocutaneous sensory augmentation affords more precise shoulder position command generation for control of FNS orthoses. In: Proceedings of the annual conference on rehabilitation technology, pp 228–230

Rohland TA (1975) Sensory feedback for powered limb prostheses. Med Biolo Eng Comput 13(2):300–301

Schorr SB, Quek ZF, Romano RY, Nisky I, Provancher WR, Okamura AM (2013) Sensory substitution via cutaneous skin stretch feedback. IEEE international conference on robotics and automation (ICRA), pp 2333–2338

Scott RN, Brittain RH, Caldwell RR, Cameron AB, Dunfield VA (1980) Sensory-feedback system compatible with myoelectric control. Med Biolo Eng Comput 18:65–69

Shannon GF (1976) A comparison of alernative means of providing sensory feedback on upper limb prostheses. Med Biolo Eng 14:289

Srinivasan MA, Beauregard GL, Brock DL (1996) The impact of visual information on the haptic perception of stiffness in virtual environments. In Proceedings of the 5th international symposium on haptic interfaces for virtual environment and teleoperator systems, american society of mechanical engineers dynamic systems and control division, vol. 58, pp 555–559

Srinivasan MA, LaMotte RH (1995) Tactual discrimination of softnesss. J Neurophysiol 73(1):88–101

Stanley AA, Kuchenbecker KJ (2011) Design of body-grounded tactile actuators for playback of human physical contact. In Proceedings of the ieee world haptics conference, pp 563–568

Tan HZ, Pang XD, Durlach NI (1992) Manual resolution of length, force, and compliance. In: Proceedings of the 1st international symposium on haptic interfaces for virtual environment and teleoperator systems, American society of mechanical engineers dynamic systems and control division, vol 42, pp 13–18

Tan HZ, Srinivasan MA, Reed CM, Durlach NI (2007) Discrimination and identification of finger joint-angle position using active motion. ACM Trans Appl Percept 4(2):10

Taylor JL, McCloskey DI (1992) Detection of slow movements imposed at the elbow during active flexion in man. J Physiol 457:503–513

Thakur PH, Bastian AJ, Hsiao SS (2008) Multidigit movement synergies of the human hand in an unconstrained haptic exploration task. J Neurosci 28(6):1271–1281

Tiest WMB, Kappers AML (2009) Cues for haptic perception of compliance. IEEE Trans Haptics 2(4):189–199

Varadharajan V, Klatzky R, Unger B, Swendsen R, Hollis R (2008) Haptic rendering and psychophysical evaluation of a virtual three-dimensional helical spring. In Proceedings of the 16th international symposium on haptic interfaces for virtual environments and teleoperator systems, pp 57–64

Weinstein S (1968) Intensive and extensive aspects of tactile sensitivity as a function of body part, sex, and laterality. In: Kenshalo DR (ed), (chap. 10)

Wheeler J, Bark K, Savall J, Cutkosky M (2009) Investigation of rotational skin stretch for proprioceptive feedback with application to myoelectric prostheses. IEEE Trans Neural Syst Rehabili Eng 18(1):58–66

Witteveen HJ, Luft F, Rietman JS, Veltink, PH (2013) Stiffness feedback for myoelectric forearm prostheses using vibrotactile stimulation. IEEE Transactions on Neural Systems and Rehabilitation Engineering, EPub ahead of print

Chapter 11
A Fabric-Based Approach for Softness Rendering

Matteo Bianchi, Alessandro Serio, Enzo Pasquale Scilingo and Antonio Bicchi

11.1 Introduction

The reproduction of material properties like *softness* is a crucial component for a compelling and realistic experience of tactile interaction. Softness is a property specifically related to tactile information and hence to the semantic representation of objects (Klatzky et al. 1991; Lederman and Klatzky 1987; Newman et al. 2005). Indeed, tactile signals about softness are among the most accessible sources of information after the initial phases of contact (Lederman and Klatzky 1997b) since their coding does not require any geometrical description of the object (Klatzky et al. 1989). Softness perception relies on two types of sensory signals: cutaneous (tactile) information (which is mainly related to the mechanical deformation of the skin) and proprioception/kinaesthesia [which can be regarded as the internal sensing of forces, displacements and postures processed inside joints, muscles, tendons and skin (Bastian 1888)], even though other modalities could also contribute to some extent like vision and audition, see Chaps. 2 and 4 respectively. Both types of information are necessary to have a softness perception for compliant objects with rigid surfaces. However, during normal interaction with the world, tactile sensory information is predominant, as the cutaneous contribution alone is sufficient for softness discrimination of objects with deformable surfaces (Srinivasan and LaMotte 1995). Most

M. Bianchi (✉) · A. Bicchi
Department of Advanced Robotics (ADVR), Istituto Italiano di Tecnologia,
via Morego 30, 16123 Genoa, Italy
e-mail: matteo.bianchi@iit.it; matteo.bianchi@centropiaggio.unipi.it

A. Bicchi
e-mail: bicchi@centropiaggio.unipi.it; antonio.bicchi@iit.it

M. Bianchi · A. Serio · E.P. Scilingo · A. Bicchi
Research Centre "E. Piaggio", Università di Pisa,
Largo Lucio Lazzarino 1, 56126 Pisa, Italy

A. Serio
e-mail: alessandro.serio@centropiaggio.unipi.it

E.P. Scilingo
e-mail: e.scilingo@centropiaggio.unipi.it

© Springer-Verlag London 2014
M. Di Luca (ed.), *Multisensory Softness*, Springer Series on Touch and Haptic Systems,
DOI 10.1007/978-1-4471-6533-0_11

219

haptic devices which are currently available (Hannaford and Okamura 2008) act primarily as force displays, although a cutaneous sensation is nevertheless provided through the contact with the device tool, but it is not modulated by the device. By contrast, tactile displays stimulate skin by conveying force with both contact and shape information. To convey cutaneous information with these devices, it is necessary to reproduce on the finger pad the complex mechanical interaction and stress/strain distribution which originates from the contact between the finger and the external object. Such tactile displays stimulate the mechanoreceptors that basically react to the strains of the skin in a manner proportional to the velocity, acceleration, or elongation (Kern 2009). Trying to measure and reproduce the tensor distribution produced by the human skin (itself a dishomogeneous, anelastic material), is a difficult task. The challenge for research is to reduce the complexity of the tactile information to a meaningful approximation, while considering design limitations such as feasibility, costs and quality of the rendering of the haptic stimuli.

11.2 Taming the Complexity of Haptic Information

To identify models which harness the complexity of tactile sensing, Bianchi (2012) proposed a geometrical reduction method, mapping a high dimensional space of perceptual elemental variables (such as information provided by sensory receptors) to a low dimensional space comprised of perceptual primitives and performance variables. The goal was to use these primitives to drive the design of haptic devices and artificial systems, which might enable a more reliable human-machine interaction. This approach draws inspiration from the neuroscientific studies on the biomechanical and neural apparatus of the human hand, which demonstrate that, despite the hands complexity, the simultaneous motion and force of the fingers is characterised by coordination patterns that reduce the number of independent Degrees of Freedom (DoFs) to be controlled (Schieber and Santello 2004). This experimental evidence, which can be explained in terms of central and peripheral constraints in the neuromuscular apparatus, describes well the concept of *hand synergies* (synergy, from Greek *work together*), i.e. the aforementioned covariation schemes observed in digit movements and contact forces. For a complete review on the concept of synergies (see Santello 2013).

A valuable conclusion can be drawn from these results: for a wide range of hand behaviours the kinematic space of the hand has a smaller dimensionality than the one represented by its mechanical degrees of freedom. Synergies can be regarded as *maps* (Latash 2008) between the higher dimensional complexity of purely mechanical architecture of the human hand and the lower dimensional control space of the action and performance. In the latter it is possible to individuate kinematic and kinetic primitives. In robotics, the concept of synergies has been used to design robotic hands in a simplified manner, and has lead to promising initial results [from the early attempts, e.g. (Brown and Asada 2007; Ciocarlie et al. 2007; Ciocarlie and Allen 2009), to the most recent applications (Catalano et al. 2012)]. Furthermore, synergies were also used to guarantee optimal performance and design for glove-based Hand Pose Reconstruction (HPR) systems (Bianchi et al. 2013a, b).

As previously mentioned, Bianchi (2012) extended the aforementioned motoric concepts to the (dual) haptic sensing domain, to find a mapping between the higher-dimensional redundant space of elemental sensory variables involved in the mechanics of touch and the lower dimensional space of perceptual primitives, i.e. "what we actually feel". An attempt to accomplish this goal was pioneered in Bicchi et al. (2011), where authors hypothesised the existence of a sort of *sensory synergy* basis. The elements of this basis can be regarded as tactual perception manifold projections onto constrained subspaces, where the subspaces increasingly individuate refined approximations of the full spectrum of haptic information. Hayward (2011) investigated the existence of a *plenhaptic function* for determining the dimensionality of haptic perception. This function is the haptic counterpart of the plenoptic function (Adelson and Bergen 1991), defined in the visual domain to indicate the number of coordinates necessary to describe all possible sensorimotor interactions. From a mechanical point of view, the plenhaptic function can be regarded as the complete characterisation of haptic experience. In terms of vector basis it comprises all the elements necessary for an exhaustive description. In Hayward (2011) it was also noted that even if the number of dimensions needed to describe mechanical interactions in haptics is larger than three or four, human touch-related experience seems to take place in a lower dimensional space; i.e. the nervous system produces nearly instantaneous reductions of dimensions, to convert a complex problem into a manageable set of computational tasks. Tactile illusions, for example, can be interpreted as the results of these low dimensional simplifications of the plenhaptic function—sampled in time and space—related to motoric and sensory capabilities.

11.3 The Contact Area Spread Rate Approach for Softness Rendering

In softness discrimination, a possible reduction of the dynamic, force-varying tactile information operated by the nervous system might be described by the *tactile flow* paradigm (Bicchi et al. 2005, 2008), which extends Horn and Schunk's equation (Horn and Schunk 1981) for image brightness to three-dimensional strain tensor distributions. Tactile flow equation suggests that, in dynamic conditions, a large part of contact sensing on the finger pad can be described by the flow of Strain Energy Density (SED) (or Equivalent von Mises Stress), since Merkel-SA1 afferents, which are primarily responsible for dynamic form in tactile scanning, were proven to be selectively sensitive to these scalar quantities (Johnson 2001). Moreover, the integral version of the tactile flow equation can be used to explain the *Contact Area Spread Rate (CASR)* (Bicchi et al. 2000) experimental observation, which affirms that a considerable part of tactile ability in object softness discrimination is retained in the relationship between the contact area growth over an indenting probe (e.g. the finger

Fig. 11.1 The discrete CASR display (**a**). A finger interacting with the discrete CASR device (**b**)

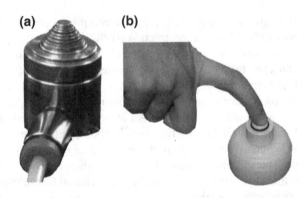

pad that presses the object) and the indenting force itself. These suitable approximations and reductions of the haptic information manifold suggest new strategies for building haptic interfaces. For example, recognizing that a simple force-area relation describes a large amount of the cutaneous information involved in softness discrimination by probing has inspired the development of simpler and more effective softness displays for human-machine interaction.

11.3.1 The First Discrete CASR-Based Display

Bicchi et al. (2000) presented the first prototype of a CASR-based display. Its role is to mimic the rate at which the contact area of the probed material grows over the surface of the probing finger pad. The implementation proposed in Bicchi et al. (2000) consists of a set of cylinders of different radii, assembled in telescopic arrangement (see Fig. 11.1).

As a result of the discontinuity in the structure due to the cylinders, this CASR display will be referred to as the *discrete CASR display*. Regulated air pressure acts on one end of the cylinders according to the desired force to be perceived by users during the indentation (see Fig. 11.1). Pressure is applied on all the cylinders. When the user finger probes the display, pushing down against the cylinders, it comes into contact with a surface depending on the height of the cylinders themselves and perceives a resultant force correlated to the pressure. The length of the cylinders is chosen such that, when the operator exerts no force, the active surface of the display can be approximated in a stepwise manner as a cone whose vertex has a total angle of $2a$. After the finger is pushed down by an amount of δ, the contact area A can be approximately computed as $A(\delta) = \pi \, \delta^2 \, \tan(\delta)^2$. Consequently, the resultant force F that is opposed to the finger is $F(\delta) = pA(\delta)$, where p is the pressure to be provided to the inner chamber of the device by the external pressure regulator. In this prototype, the displacement δ can be measured with an optoelectronic sensor or a proportional Hall sensor placed at the bottom of the inner chamber. In Bicchi et al. (2000) psychophysical experiments proved that the discrete CASR display is able to provide better performance in softness discrimination than the one achieved using a purely kinaesthetic display (i.e. the discrete CASR display covered with a hollow rigid cylinder).

The experimental curves F/A (or CASR curves) obtained for the discrete CASR display, at fixed pressure levels, are linear; to mimic CASR curves of real objects, typically nonlinear, the display has to be controlled acting on the variable p.

11.4 Fabric-Based Displays

The discrete CASR display was proven to be able to replicate desired force-area curves and to enable a more realistic softness perception compared with the one achievable with a purely kinaesthetic device. However, the structure of this display does not provide users with a continuously deformable surface, thus producing edge effects. This fact might lead to a not completely immersive experience, which can destructively affect the transparency and reliability of the perception, e.g. in tele-operation tasks. Moreover, the contact area involved in the interaction can be known only after some geometric considerations related to the measured displacement. This can represent a limitation for correctly mimicking real CASR curves, for which a real-time accurate measurement of the contact region is mandatory.

To overcome these limits, we propose a new concept of displays based on a bi-elastic fabric, hereinafter defined as Fabric Yielding Displays (FYDs). Bi-elastic means that the fabric exhibits properties that render it elastic in at least two substantially perpendicular directions. After preliminary tests on different materials, we decided to use the Superbiflex by Mectex (Erba, Como, Italy) since it offers both good elastic behaviour within a large range of elasticity, and a high resistance to traction. By changing the elasticity of the fabric, users are able to feel different levels of softness by touching a deformable surface. At the same time the contact area on the finger pad can be measured via an optical system.

11.4.1 Introduction to the First FYD Prototype

The first prototype of the FYD consists of a hollow plastic cylinder (ABS 3D printed, 195 × 50 mm) containing a linear actuator (Linear Actuator L16-100-35-12-P by Firgelli, Victoria BC, Canada), which is a compact DC motor geared to push or pull loads along the stroke (100 mm). This actuator comes with a built-in potentiometer and control (maximum positional error of 0.4 mm), allowing one to monitor and set the actuator position. On the top of the cylinder a rectangular shaped piece of fabric (200 × 200 mm) is placed and tied to a circular crown. The crown is attached to the motor stroke and it can run alongside the cylinder with minimal friction. When the motor pulls down the crown, the fabric is stretched and its apparent stiffness

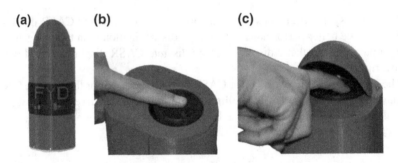

Fig. 11.2 The FYD first prototype, an overview (**a**). A finger interacting with the display. For the sake of clarity the FYD is shown without and with the cover in (**b**) and in (**c**), respectively

increases. Conversely, when the motor pushes up the crown, the fabric is relaxed and feels softer (Bianchi et al. 2010). The linear actuator is able to move at a rate of 32 mm/s and can exert up to 50 N of force at lower speeds, for a range of motion of 30 mm. The FYD also behaves like a contact area display. A web camera (Hercules web cam, resolution of 320 × 240 pixels at 30 frames/s) is placed inside the hollow cylinder, at the centre of the mechanical interface, just beneath the fabric (at a distance of ≃ 300 mm). The camera is equipped with high luminosity Light Emission Diodes (LEDs) and frames the lower surface of the fabric, in particular the image of the strained fabric after the indentation. During tactual probing, the fabric is strained and the fabric area in contact with the fingertip changes in proportion to the applied force. The contact area can be estimated using the algorithms described below. The prototype as a whole is connected to a base and enclosed within a protective shell, with total dimensions of 195 × 115 × 115 mm. An overview of the system and an exploded drawing view are reported in Figs. 11.2 and 11.3, respectively. For further details on the architecture of the first FYD prototype the reader can refer to (Bianchi et al. 2009, 2010).

11.4.2 Area Acquisition

The contact area between the fabric and the finger can be estimated and visually displayed by means of a suitable segmentation algorithm. The contact area acquisition algorithm is based on RGB image binarisation. More properly, only one image band (the R band, which is a 320 × 240 matrix of integer numbers) out of three is involved in the area detection algorithm to reduce the computational workload and allow for fast processing. The underlying idea is quite simple; while the fabric is probed, the indented fabric surface is closer to the camera with respect to the outer region, see Fig. 11.4. Consequently, this area will be more brightly lit by the LEDs. The difference between background luminosity and contact area luminosity is discriminated via binarisation thresholds, heuristically calculated. Using a linear

Fig. 11.3 The FYD first prototype: exploded drawing view

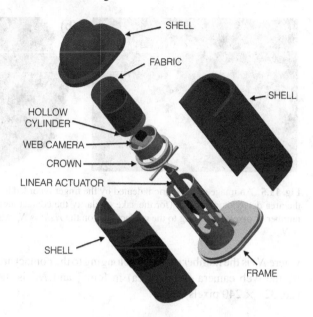

Fig. 11.4 The measurement of the contact area

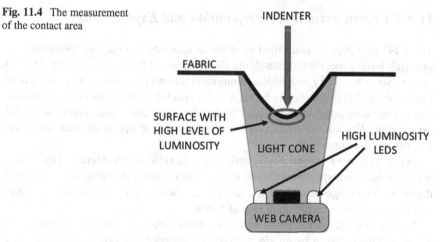

interpolation, a binarisation threshold is associated to each vertical position of the crown. In this manner, the pixels in the image that belong to the contact area can be individuated and displayed (see also Fig. 11.5). In order to guarantee uniform and repeatable luminosity conditions, a cover is placed on the top of the external shell of the device (see Fig. 11.2). The contact area $A_{contact}$ expressed in (cm^2) is estimated as

$$A_{contact} = N_c \times \frac{A_f}{N_w}, \tag{11.1}$$

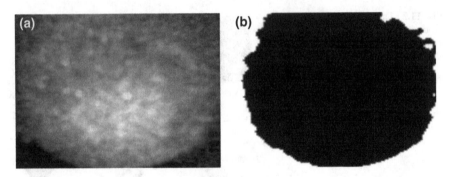

Fig. 11.5 An image of the fabric indented by the finger recorded by the camera (**a**). The result of the area detection algorithm (for the sake of clarity the contact area is shown in *black*) (**b**). The number of pixels belonging to the contact area on the *right* is N_c while the total number of pixels is N_w

where N_c is the number of pixels belonging to the contact area; A_f is the area captured by the web camera (frame area) in [cm^2] and N_w is the web camera resolution (i.e. 320×240 pixels).

11.4.3 Characterisation, Interpolation and Experiments

The FYD prototype is controlled in order to simulate mechanical compliance of materials having specific (stiffness) force/displacement ($F(\delta)$) and force/area ($F(A)$) curves. Since force (F) and displacement/indentation (δ) are the primary objects of kinaesthesia, $F(\delta)$ profiles can be regarded as useful abstractions and approximations of the kinaesthetic behaviour of materials; at the same time, based on CASR assumption, $F(A)$ curve contains a large part of the cutaneous information useful for softness discrimination.

In Fig. 11.6, the $F(\delta)$ and $F(A)$ curves of the fabric at different levels of stretching are reported. More specifically these levels were obtained changing the position of the crown, in a range between 0 mm (0 mm was chosen near the top of the cylinder) and 30 mm, with an incremental step of 5 mm.

For the characterisation phase we used a compressional rigid indenter shaped as a human finger, driven by an electromagnetic actuator and capable of applying a maximum controlled displacement of 10 mm in the axial direction, while the contact area, the displacement(indentation) and the contact force were measured, by the web camera, a magnetic linear transducer and a load cell, respectively. The shape of the wood indenter—15 mm in diameter and 100 mm in length—represents a first geometrical approximation of the human fingertip (Dandekar et al. 2003). The load-cell was mounted on the indenter to measure the force applied on the fabric during the indentation. What is noticeable is that $F(A)$ and $F(\delta)$ curves are linear over all the positions of the crown. During the characterisation phase, only a finite set of positions was acquired. For intermediate values, a piecewise linear interpolant was adopted. It is possible to mimic a given material with a specific stiffness coefficient (which can be

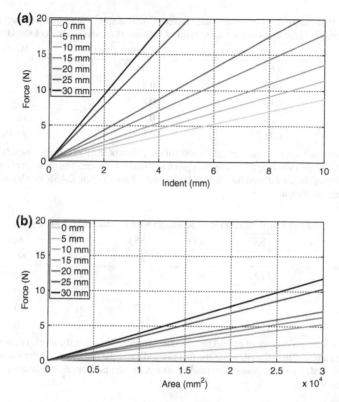

Fig. 11.6 Force/indentation characteristics at different fixed positions of the crown (**a**). The indentation is measured using a magnetic linear transducer. Force/area characteristics at different fixed positions of the crown (**b**). Notice that the position at 0 mm is close to the *top* of the external cylinder

regarded as the angular coefficient of the $F(\delta)$ linear curve), by suitably identifying the corresponding position of the crown. Moreover, from the actual measurement of the contact area, since $F(\delta)$ and $F(A)$ curves are coupled, an indirect estimation of the indented force and hence of the displacement can be obtained. Notice that these estimations are strongly related to the nature of the contact. Indeed, in order to obtain coherent values, the user should touch the fabric in the same manner as the wood indenter did during the characterisation phase.

11.4.4 Evaluation Experiments

We designed an experimental session to evaluate the performance of the FYD, comparatively with the discrete CASR display, using five simulated levels of stiffness ($SS1$, $SS2$, $SS3$, $SS4$ and $SS5$). Table 11.1 shows the input parameters for the FYD and for the discrete CASR device necessary to reproduce the five simulated stiffness values.

Table 11.1 Parameters used to control the discrete CASR display (third column) and the FYD (fourth column) in order to render the same level of stiffness (first and second columns)

Stiffness	Coeff. (N/cm)	Pressure (bar)	Position (mm)
SS1	0.67	0.35	1.6
SS2	1.00	0.5	4.1
SS3	1.18	0.6	8.6
SS4	1.28	0.7	14.4
SS5	1.71	0.8	23.3

The term "Position" refers to the vertical position of the crown of the FYD (the position 0 is chosen close to the top of the cylinder) associated to a given stiffness coefficient. The term "Pressure" refers to the pressure of the air inflated into the inner chamber of the discrete CASR display to mimic a given stiffness coefficient

Table 11.2 Confusion matrix of ranking experiments with the discrete CASR display

	SS1	SS2	SS3	SS4	SS5	Accuracy (%)
SS1	18	4	1	1	6	60
SS2	0	19	8	3	0	63
SS3	2	1	18	8	1	60
SS4	8	1	1	17	3	56
SS5	2	5	2	1	20	66

The accuracy is the percentage of correct recognition, associated to a specific level. The first column contains the names of the stimuli, while the first raw indicates the responses, i.e. how stimuli were identified (ranked) by participants. The total accuracy, i.e. the percentage of correct recognition across all levels, is 61 %

After providing written consent, 10 right-handed volunteers participated in the study (7 males and 3 females, their age ranged from 23 to 40) (Bianchi et al. 2010). None had a history of limitations that could affect experimental outcomes. They performed the tests blindfolded and with ear plugs, to prevent the possible use of any other sensory cues and eliminate any diversion from the task. They were presented with different levels of stiffness and asked to judge them by pressing vertically or tapping their index finger against the displays. In ranking tests, which were conducted independently with each device, participants were presented with new stimuli for less than one second and each trial was repeated three times per participant. They were asked to probe the set of five levels $SS1$ to $SS5$, presented in random order, and sort in terms of softness,. Participants were asked to rank the stimuli from 1 to 5, where 1 corresponded to the softest.

Results from ranking experiments are shown in Tables 11.2 and 11.3, where subjective softness is reported versus objective compliance in a confusion matrix structure (Srinivasan and LaMotte 1995) for the five levels, using both the devices. Values on the diagonal express the number of correct answers. The percentage of total accuracy is calculated considering the sum of all correct answers for all the levels of stiffness. The correspondence between an objective estimation of the compliance and the subjective evaluation in terms of numerical values in a given scale was already

Table 11.3 Confusion matrix of ranking experiments with the FYD

	SS1	SS2	SS3	SS4	SS5	Accuracy (%)
SS1	22	4	0	2	2	73
SS2	4	25	1	0	0	83
SS3	0	1	27	0	2	90
SS4	3	0	0	25	2	83
SS5	1	0	2	3	24	80

The accuracy is the percentage of correct recognition, associated with each specific level. The first column contains the names of the stimuli, while the first raw indicates how they were identified (ranked) by participants. The percentage of correct recognition across all levels, is 82 %

used e.g. in Srinivasan and LaMotte (1995), and Friedman et al. (2008). The results obtained with the discrete CASR display exhibits a percentage of total accuracy of 61 %, while with the FYD the percentage of total accuracy is 82 %. Notice that the chance level is 20 %.

Results show that the FYD enables a better softness perception than the discrete CASR display. This enhancement is probably due to the absence of edge effects during the interaction between the fingertip and the fabric surface; indeed, the FYD provides cues for a more reliable and realistic perception, since the fabric is deformable in a controlled manner under the finger pad. This fact might also help to develop in a more effective manner the haptic memory required for multiple comparisons.

11.5 The Second Version of the FYD: The FYD-2

Although the first FYD prototype was proven to be able to enhance softness discrimination accuracy in participants, by conveying contact area cues in an intuitive and efficient manner, there were still some design aspects to be improved. Indeed, although the contact area was actively measured, no contact-area feedback for dynamic tracking was implemented. Furthermore, the physical dimensions of the device can potentially prevent it from integration in multi-device systems such as in Scilingo et al. (2010), for tasks where space constraints are mandatory. For these reasons, we created a second version of the display, hereinafter referred to as FYD-2 (Serio et al. 2013) (cfr. Fig. 11.7). The main advantages of the FYD-2 design are: the reduced dimensions (70 × 70 × 100 mm), which enable possible integrations with other devices and wearability; an actuation system based on two fast motors and a more effective sensorisation scheme, which consists of a web camera, for real-time measurement, and a force sensor mounted at the base of the device, to record the normal contact force exchanged between the finger pad and the fabric. Other approaches found in literature lack real-time area measurement, and this severely limits the reliability of tracking $F(A)$ curves by introducing edge effects and discretisation (Bicchi et al. 2000) or allowing the control of the fingertip contact area

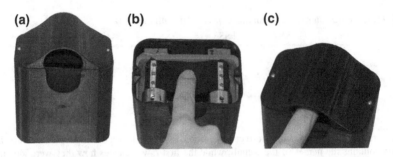

Fig. 11.7 The FYD-2, an overview (**a**). A finger interacting with the display. For the sake of clarity the FYD-2 is shown without and with the cover in (**b**) and in (**c**), respectively

only for a finite set of constructed and stored in advance "numerical models" as in Fujita and Ohmori (2001). Here, the proposed actuation and sensorisation scheme realises a closed-loop control, based on the actual measurement of the contact area, which allows to track arbitrary force/area characteristics.

Furthermore, the actuation scheme endows the system with an additional degree of freedom, which can be used to convey supplementary haptic cues, such as directional information, for a more compelling and immersive haptic experience.

11.5.1 Mechanical Design

In the second version of the device the extremities of a rectangular strip of the fabric are connected to two rollers. These rollers are independently moved by two DC Maxon Motors REmax—256 : 1, 3 Watt—(Maxon Motor ag, Sachseln, Switzerland) through two pulleys placed on motor shafts.

Motor positions can be controlled by processing the signals from two absolute magnetic encoders (12 bit magnetic encoder by Austria Microsystems—Unterprem-staetten, Austria—AS5045 with a resolution of 0.0875°), read by a custom made electronic board (PSoC-based electronic board with RS–485 communication protocol).

A level of softness is generated by appropriately stretching the fabric using the two motors; i.e. when motor 1 rotates in a counter-clockwise direction and motor 2 rotates in a clockwise direction they stretch the fabric thus increasing its apparent stiffness. When motor 1 rotates in a clockwise direction and motor 2 rotates in a counter-clockwise direction they relax the fabric thus reducing its apparent stiffness, see Fig. 11.8. Furthermore, it is important to notice that the two-motor configuration allows one to implement and exploit an additional degree of freedom. Indeed, when the two motors coherently rotate in the same direction, a translational shift can be imposed on the finger pad interacting with the fabric, as it is shown in Fig. 11.8. This shift can be used to convey kinaesthetic, directional and vibrotactile information

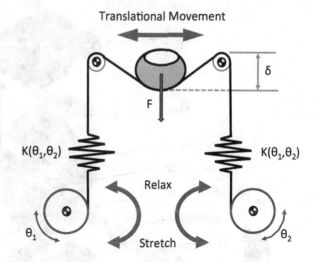

Fig. 11.8 How the system works during the interaction with a human finger pad. K is the stiffness of the fabric, which depends on motor positions, θ_1 and θ_2

to users. Finally, since motors can be independently controlled, it is also possible to modify contact area geometry (eccentricity), thus simulating incipient slippage conditions or curvature perception.

The FYD-2 is also endowed with a load cell (Micro Load Cell (0–780g) - CZL616C from Phidgets, Calgary, Alberta, Canada) placed at the base of the device, to record the normal force exerted by the user finger interacting with the fabric. Notice that the shear force is not considered. This is in agreement with the instructions given to users interacting with the device. Indeed they were recommended to not perform movements of the finger across the surface and to not apply lateral forces, in order to eliminate any anisotropic effect or distortion in softness but only focusing on normal indentation of the specimens (Lederman and Klatzky 1997a).

The system has also a web camera (Microsoft "LifeCam HD–3000" with a resolution of 640 × 480) and two high luminosity LEDs (whose luminosity can be regulated with a trimmer) just beneath the fabric (30 mm), for measuring the contact area in real-time (cfr. Fig. 11.9). Force and area information are then used to implement the force/area tracking algorithm, whose results are described in the following subsections.

11.5.2 Experiments

As discussed in the Introduction, softness perception relies on both haptic channels—kinaesthesia and cutaneous information—although tactile sensing plays a predominant role. Based on these considerations, the fabric-based device exhibits two types of behaviour and hence of control:

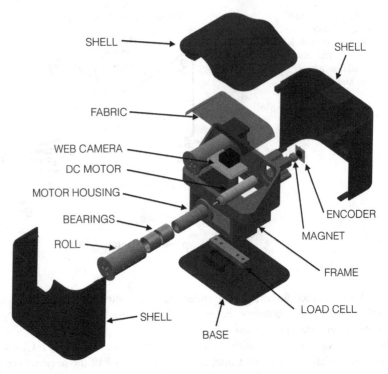

Fig. 11.9 FYD-2, exploded drawing view

(I) the FYD-2 can be controlled to track $F(\delta)$, while acting as a contact area (real-time) display

(II) the FYD-2 can be controlled to track $F(A)$, while actively using, together with the measured force, the real-time measured contact area as a feedback signal.

Notice that both $F(\delta)$ and $F(A)$ curves are not independent since they represent the cutaneous and kinaesthetic characterisation of the softness properties of a given object. Therefore they are determined by the fabric (object) characteristics. This represents a common limitation for all haptic devices: to properly decouple these curves it would be necessary to have two independent control variables, as proposed for example by Scilingo et al. (2010), where a conventional kinaesthetic haptic display is combined with a cutaneous softness one.

For case (I), the aim is to mimic a given stiffness. In this case, the display behaves like common kinaesthetic devices (Hannaford and Okamura 2008 which can be regarded as force displays), although cutaneous cues and the measurement of the contact area provide additional information. On the other side, case (II)—$F(A)$ tracking problem—represents a more challenging issue, given that the law that relates the growth of the contact area and the indenting force, i.e. the CASR paradigm, represents a large part of the tactile information used for softness discrimination.

To implement these controls, the first step is to characterise the device. The characterisation procedure is analogous to the one used for the first prototype. Different positions of the motors were considered, with same starting point and angular displacement used for both. The angular values ($\theta = \theta_1 = -\theta_2$) used for the characterisation range from $10°$ to $80°$, with an incremental step of $10°$. The range of contact force is from 0 to 20 N. In this case, the force/displacement characteristics, interpolated at fixed motor positions, are quadratic ($R^2 > 0.94$), and the stiffness [σ in (N/mm)] of the fabric can be computed directly deriving the contact force with respect to the displacement. Let the characteristic be $F = \lambda\delta^2$, with λ (N/mm^2) representing the quadratic coefficient of the parabolic curve at fixed motor positions. In this case the stiffness of the fabric depends on the displacement and it can be defined as (Grioli and Bicchi 2010; Serio et al. 2011): $\sigma(\delta) = \partial F/\partial\delta = \rho\delta$, where $\rho = 2\lambda$ (N/mm^2) represents the stiffness coefficient.

11.5.2.1 Constant Stiffness Tracking

For this kind of experiment, we used the finger pad of a right-handed male participant (age 32) as the indenter for probing the fabric approximately every second. Since the fabric stiffness is not constant but it depends on the indentation, we control motor positions using motor encoders to know θ value. From this value it is then possible to retrieve the angular coefficient [ρ_a in (N/mm^2)] of the actual stiffness curve from the characterisation characteristics or interpolating between them. Using the information about the contact force measured by the load cell of the device, the actual indentation δ can be obtained as $\delta = \sqrt{F/\lambda}$. Finally, the actual stiffness [σ_a in (N/mm)] of the fabric can be computed as $\sigma_a = \rho_a\delta$ (see also Fig. 11.10 for more details).

In Fig. 11.11, the results of tracking a constant stiffness of $\sigma_r = 1\ N/mm$ and the control scheme adopted are reported. A PID control (constants $P = 1, I = 0.01$, $D = 0$, heuristically found) is then used to control motor positions, based on the error (e) between σ_r (reference stiffness) and σ_a (actual stiffness). In this case, after an initial transitory phase due to motor positioning, we get an RMSE of $0.18\ N/mm$, less than 20 % of the reference value. The effect of the transitory phase on human perception will be investigated in future experiments.

11.5.2.2 Trajectory Area Tracking

In order to reproduce common quadratic $F(A)$ characteristics (Bicchi et al. 2000), the position of the motors needs to be controlled and suitably rapidly changed, based on the actual contact area. This fact motivated the need for a fast actuation system. Let be $F(A) = \xi_r A_r^2$, the quadratic curve to be tracked, with ξ_r in (N/mm^4) the quadratic coefficient of the curve. In order to properly implement the control, we need to know the actual ξ_a (N/mm^4) coefficient. This coefficient is obtained each time by dividing the indenting force measured by the load cell for the squared value of the measured area A_m. A PID controller is then used, which is based on the error between ξ_r and the

Fig. 11.10 The control
scheme used for constant
stiffness tracking. A PID
control is used to control
motor positions, based on the
error (*e*) between σ_r (reference stiffness) and σ_a (actual
stiffness)

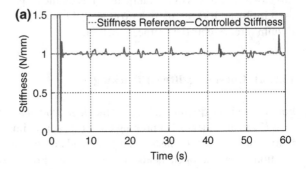

Fig. 11.11 a Stiffness
control. **b** Force.
Reference stiffness
(*dashed line*) versus
controlled stiffness
(*continuous line*) (**a**). Force
measurement (**b**)

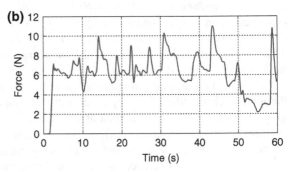

actual ξ_a (see Fig. 11.12), constants: $P = 5, I = 0.3, D = 0$. The effectiveness of this
control scheme was experimentally verified by tracking the characteristic curves of
different silicone specimens, realised with different percentage of plasticiser, whose
characteristic curves were experimentally obtained, as described below.

For the sake of space, in Fig. 11.13 we report only the results for the specimen
at 0 % plasticiser percentage. In this case, we get an RMSE of 41.3 mm^2, less than
14 % with respect to the reference value.

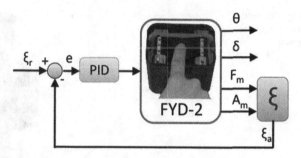

Fig. 11.12 The block diagram of control for $F(A)$ curve tracking. Let be $F(A) = \xi_r A_r^2$, the quadratic curve to be tracked, with ξ_r in (N/mm^4) the quadratic coefficient of the curve. In order to properly implement the control, we need to know the actual ξ_a (N/mm^4) coefficient. This coefficient is obtained each time by dividing the indenting force measured by the load cell for the squared value of the measured area A_m. F_m is the measured force. A PID controller is then used to control the quadratic coefficient

Fig. 11.13   a $F(A)$ Control.
b Force. Reference area
(*dashed line*) versus
Controlled area (*continuous line*) (**a**). Force measurement
(**b**)

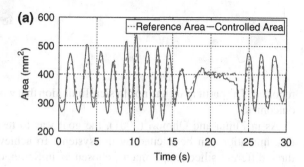

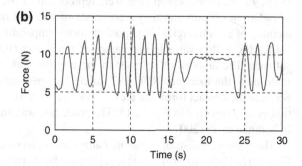

11.5.3 Evaluation Experiments

We report a preliminary assessment of the FYD-2 performance by a comparative evaluation of the rendered softness and the objective compliance of real materials. Results of such experiments from one right-handed male participant (age 27) are

Fig. 11.14 The system used
for the characterisation of the
silicone specimens

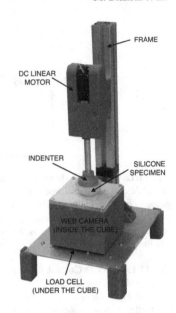

reported. The participant had no physical limitation that would affect the experimental
outcomes.

As in Fujita and Ohmori (2001), the goal was to test how effectively softness
discrimination can be elicited by the system. To achieve this objective, we used
three different silicone specimens, chosen as in Scilingo et al. (2010). The specimens, whose softness properties were reproduced by the FYD-2, were half–spheres
of radius of 20 mm and they were made of material obtained by mixing a given
quantity of a commercial bicomponent, room temperature-curing silicone (BJB TC-
5005A/B), with a variable percentage of plasticiser (BJB TC-5005C), acting as a
softener. The amount of softener in the mixture was chosen as 0, 10 and 20 %,
referring to the specimen $SS1$, $SS2$ and $SS3$, respectively. To derive the relationship
between the contact force and the contact area we used a custom made characterisation system reported in Fig. 11.14. The procedure was analogous to the one reported
in Scilingo et al. (2007).

The system consists of an indenter attached to a servo-controlled linear actuator by
Firgelli (L-12-50-100-6-L). The specimens to be characterised were placed under the
motor stroke and put in progressive contact with a transparent glass. The indenting
velocity was 5 mm/s and the range of force was $0 \div 20$ N.

A webcamera—Microsoft LifeCam HD-5000—was put under the glass. As the
indenter pushed against the specimen the web cam captured a snapshot of the surface
flattened against the plexiglass. In order to enhance contours of contact area a thin
white paper behaving as optical filter was placed between the specimen and the
plexiglass.

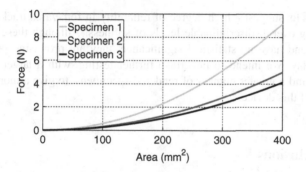

Fig. 11.15 The force/area ($F(A)$) characteristics for the silicone specimens

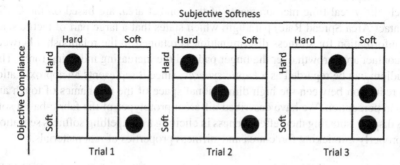

Fig. 11.16 Confusion matrices showing how the objective compliance was subjectively perceived by one participant

Following the previously described techniques, we obtained the contact area on the basis of heuristically found binarisation thresholds based on luminosity. Furthermore, for each contact area, the indentation force was also measured by means of a load cell placed at the base of the system. In this manner the $F(A)$ curves were obtained for $SS1$, $SS2$ and $SS3$, see Fig. 11.15.

Using the previously adopted terms, let be $SH1$, $SH2$ and $SH3$ the rendered stimuli corresponding to $SS1$, $SS2$ and $SS3$, respectively. The rendered stimuli were obtained by reproducing the $F(A)$ curves of the silicone specimens, using the control scheme described in the previous subsection. $SH1$, $SH2$ and $SH3$ were presented three times in a random order to the participant (right-handed, male, age 30) and then he was asked to associate them to their physical counterparts. The participant did not have time limitations since he was allowed to touch the silicone specimens and the rendered stimulus as many times as he wanted. The experiment was performed in blind conditions.

Results are shown in Fig. 11.16, where the perception of artificial specimens was associated to the perception of real ones in a confusion matrix structure. In other words, the three stimuli are reported along the row dimension, while the ranked responses are reported along columns.

These results suggest a high degree of reliability in force/area tracking as well as in eliciting overall discriminable levels of softness. Although these results are preliminary and have no statistical significance, they are promising performance for this display. We intend to undertake further testing, with a larger number of participants and a greater number of rendered specimens, to give a more complete assessment of this device.

11.6 Conclusions

In this chapter we have presented fabric-based softness displays. Such displays, which allow real-time measurements of the contact area, are based on the CASR (Contact Area Spread Rate) paradigm which states that a large part of tactile sensing information for softness discrimination is retained in the relationship between the contact area growth over the finger pad and the increasing indenting force. This paradigm can be regarded as a *haptic synergy*, since it represents an approximation and reduction between the high dimensional space of the mechanics of touch and human perception. We have described different prototypes of the fabric-based softness display, showing their effectiveness in eliciting a compelling softness sensation, by properly mimicking force-area and stiffness properties of real materials.

Acknowledgments This work is supported by the European Research Council under the ERC Advanced Grant $n°$ 291166 SoftHands (A Theory of Soft Synergies for a New Generation of Artificial Hands). The research leading to these results has also received funding from the European Union Seventh Framework Programme FP7/2007–2013 under grand agreement $n°$ 248587 THE (The Hand Embodied) and under grant agreement $n°$ 601165 WEARHAP (WEARable HAPtics for humans and robots).

References

Adelson EH, Bergen JR (1991) The plenoptic function and the elements of early vision. Landy M, Movshon JA (eds) Computational models of visual processing. MIT Press, Cambridge, pp 3–20

Bastian HC (1888) The 'muscular sense': its nature and cortical localisation. Brain 10:1–137

Bianchi M (2012) On the role of haptic synergies in modelling the sense of touch and in designing artificial haptic systems. PhD thesis, University of Pisa, Pisa, Italy

Bianchi M, Salaris P, Bicchi A (2013a) Synergy-based hand pose sensing: optimal glove design. Int J Robot Res 32(4):407–424

Bianchi M, Salaris P, Bicchi A (2013b) Synergy-based hand pose sensing: reconstruction enhancement. Int J Robot Res 32(4):396–406

Bianchi M, Scilingo EP, Serio A, Bicchi A (2009) A new softness display based on bi-elastic fabric. In: World haptics conference, pp 382–383

Bianchi M, Serio A, Scilingo EP, Bicchi A (2010) A new fabric-based softness display. In: Proceedings of IEEE haptics symposium, pp 105–112

Bicchi A, De Rossi DE, Scilingo EP (2000) The role of the contact area spread rate in haptic discrimination of softness. IEEE Trans Robot Autom 16(5):496–504

Bicchi A, Gabiccini M, Santello M (2011) Modelling natural and artificial hands with sinergie. Phil Trans R Soc B 366:3153–3161

Bicchi A, Scilingo EP, Dente D, Sgambelluri N (2005) Tactile flow and haptic discrimination of softness. In: Barbagli F, Prattichizzo D, Salisbury K (eds) Multi-point interaction with real and virtual objects, pp 165–176 (STAR: Springer tracts in advanced robotics)

Bicchi A, Scilingo EP, Ricciardi E, Pietrini P (2008) Tactile flow explains haptic counterparts of common visual illusions. Brain Res Bull 75(6):737–741

Brown C, Asada H (2007) Inter-finger coordination and postural synergies in robot hands via mechanical implementation of principal component analysis. In: IEEE-RAS international conference on intelligent robots and systems, pp 2877–2882

Catalano MG, Grioli G, Serio A, Farnioli E, Piazza C, Bicchi A (2012) Adaptive synergies for a humanoid robot hand. In: IEEE-RAS international conference on humanoid robots, pp 7–14

Ciocarlie MT, Allen PK (2009) Hand posture subspaces for dexterous robotic grasping. Int J Robot Res 28(7):851–867

Ciocarlie MT, Goldfeder C, Allen PK (2007) Dimensionality reduction for hand-independent dexterous robotic grasping. In: IEEE/RSJ international conference on intelligent robots and systems, pp 3270–3275

Dandekar K, Raju BI, Srinivasan MA (2003) 3-d finite-element models of human and monkey fingertips to investigate the mechanics of tactile sense. ASME J Biomech Eng 125:682–691

Friedman RM, Hetster KD, Green BG, LaMotte RH (2008) Magnitude estimation of softness. Exp Brain Res 191(2):133–142

Fujita K, Ohmori H (2001) A new softness display interface by dynamic fingertip contact area control. In: World multiconference on systemics, cybernetics and informatics, pp 78–82

Grioli G, Bicchi A (2010) A non-invasive real-time method for measuring variable stiffness. In: Robotics science and systems

Hannaford B, Okamura AM (2008) Haptics. In: Siciliano B, Khatib O (eds) Springer handbook on robotics. Springer, Heidelberg, pp 719–739

Hayward V (2011) Is there a "plenhaptic" function? Phil Trans R Soc B 366:3115–3122

Horn BKP, Schunk BG (1981) Determining optical flow. Artif Intell 17:185–203

Johnson KO (2001) The roles and functions of cutaneous mechanoreceptors. Curr Opin Neurobiol 11(4):455–461

Kern TA (2009) Biological basics of haptic perception. Kern TA (ed) Engineering haptic devices. Springer, Heidelberg, pp 35–58

Klatzky RL, Lederman SJ, Matula DE (1991) Imagined haptic exploration in judgements of objects properties. J Exper Psychol Learn Mem Cogn 17(1):314–322

Klatzky RL, Lederman SJ, Reed C (1989) Haptic integration of object properties:texture, hardness, and planar contour. J Exper Psychol: Hum Percept Perform 15(1):45–57

Latash ML (2008) Synergy. Oxford University Press, Oxford

Lederman SJ, Klatzky RL (1987) Hand movements: a window into haptic object recognition. Cogn Psychol 19(12):342–368

Lederman SJ, Klatzky RL (1997a) Relative availability of surface and object properties during early haptic processing. J Exper Psychol: Hum Percept Perform 23(6):1680

Lederman SL, Klatzky RL (1997b) Relative availability of surface and object properties during early haptic processing. J Exper Psychol: Hum Percept Perform 23(6):1680–1707

Newman SD, Klatzky RL, Lederman SJ, Just MA (2005) Imagining material versus geometric properties of objects: an fMRI study. Cogn Brain Res 23(3):235–246

Santello M, Baud-Bovy G, Jörntell H (2013) Neural bases of hand synergies. Frontiers Comput Neurosci 7(23)

Schieber MH, Santello M (2004) Hand function: peripheral and central constraints on performance. J Appl Physiol 96(6):2293–2300

Scilingo EP, Bianchi M, Grioli G, Bicchi A (2010) Rendering softness: integration of kinaesthetic and cutaneous information in a haptic device. IEEE Trans Haptics 3(2):109–118

Scilingo EP, Sgambelluri N, Tonietti G, Bicchi A (2007) Integrating two haptic devices for performance enhancement. In: EuroHaptics conference, 2007 and symposium on haptic interfaces for virtual environment and teleoperator systems. World haptics 2007. Second Joint, IEEE, pp 139–144

Serio A, Bianchi M, Bicchi A (2013) A device for mimicking the contact force/contact area relationship of different materials with applications to softness rendering. In: IEEE/RSJ international conference on intelligent robots and systems, 2013, IROS 2013, pp 4484–4490

Serio A, Grioli G, Sardellitti I, Tsagarakis NG, Bicchi A (2011) A decoupled impedance observer for a variable stiffness robot. In: 2011 IEEE international conference on robotics and automation, pp 5548–5553

Srinivasan MA, LaMotte RH (1995) Tactile discrimination of softness. J Neurophysiol 73(1): 88–101

Chapter 12
Haptic Augmentation in Soft Tissue Interaction

Seokhee Jeon, Seungmoon Choi and Matthias Harders

12.1 Introduction

In augmented reality (AR), the interaction space exploits a real environment, and only essential virtual content is added to achieve the application's goal, thereby transforming the real space into a semi-virtual space. This procedure reduces the workload for application development to a great extent, while preserving the realism of interaction as much as possible. These benefits have enabled the adoption of AR technology in a variety of domains. One important missing component in the current AR technology is in the domain of *touch*. As will be reviewed below, the status of *haptic AR* technology is still immature, in spite of recent research endeavours. In the following, we provide a brief introduction to haptic AR, focusing on its taxonomy and associated concepts.

S. Jeon (✉)
Kyung Hee University, Seoul, Gyeonggi-do, South Korea
e-mail: jeon@khu.ac.kr

S. Choi
POSTECH, Pohang, Gyungbuk, South Korea
e-mail: choism@postech.ac.kr

M. Harders
University of Innsbruck, Innsbruck, Austria
e-mail: matthias.harders@uibk.ac.at

© Springer-Verlag London 2014

241

M. Di Luca (ed.), *Multisensory Softness*, Springer Series on Touch and Haptic Systems,
DOI 10.1007/978-1-4471-6533-0_12

12.2 Haptic Augmented Reality

12.2.1 Taxonomy and Concepts

The reality-virtuality continuum defined by Milgram and Colquhoun (1999) places *reality* and *virtuality* at the two ends of the continuum with *mixed reality (MR)* in-between, treating the degree of reality (or virtuality) as a continuous concept. Whether an environment is closer to reality or virtuality depends on the amount of information added to the environment with the computer; the more information inserted, the closer to virtuality. Based on this criterion, MR is further classified into augmented reality and augmented virtuality. In augmented reality, the main interaction occurs within the real environment, and only indispensable virtual components are implemented for the interaction (e.g., the heads-up display in an aircraft cockpit). In augmented virtuality, primary interaction occurs within a computer-generated virtual environment, and some real elements can be mixed in to improve realism (e.g., a computer game employing a virtual dancer with the face image of a real actor or actress). Despite this clear theoretical distinction, the current literature does not strictly discriminate between the two terms and rather tends to use AR and MR interchangeably. It should also be noted that the focus in literature is almost exclusively on aspects of visual augmentation.

Nonetheless, as detailed in Jeon and Choi (2009), we can define the same reality-virtuality continuum for touch and then combine it with the visual continuum. This results in the composite visual-haptic reality-virtuality continuum, as shown in Fig. 12.1. This continuum is instrumental in clarifying the associated concepts and classifying related studies, as will be seen in the next section.

12.2.2 Related Work

In the composite continuum, the left column comprises three categories of *haptic reality*: vR-hR, vMR-hR, and vV-hR, where the corresponding environments provide only real haptic sensations (hR). An interesting category here is vMR-hR, wherein a user sees mixed reality objects, but still touches only real objects. A typical example is tangible AR, where a handheld real prop is used as a tangible interface in visually mixed environments (e.g., the MagicBook in Billinghurst et al. (2001)). Another example is the projection augmented model, where a computer-generated image is projected onto a real physical object that is explored by the bare hand (Bennett and Stevens 2006). In these cases, variation of haptic properties is regarded as less important.

The three categories in the right column of Fig. 12.1, vR-hV, vMR-hV, and vV-hV, stand for *haptic virtuality* (hV). They correspond to environments with purely virtual haptic sensations. Robot-assisted motor rehabilitation is an example of vR-hV where synthetic haptic feedback is provided in a real visual environment; while an

Fig. 12.1 The extended reality-virtuality continuum for visual and haptic stimuli (Jeon and Choi 2009). Shaded areas in the continuum correspond to the realm of mixed reality

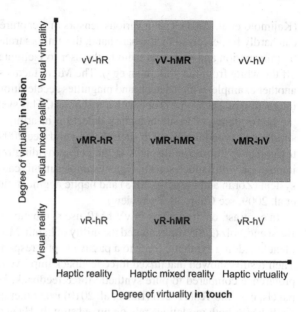

interactive virtual training simulator is an instance of vV-hV where the sensory information of both modalities is virtual. In the intermediate category, vMR-hV, purely virtual haptic objects are rendered using a haptic interface in a visually mixed environment. Required haptic rendering algorithms are not very different from the conventional algorithms for virtual objects. The central issue is the integration of virtual haptic rendering into the existing visual AR framework, with particular emphasis on precise registration between the haptic and visual coordinate frames (Vallino and Brown 1999). Bianchi et al. (2006a, b) proposed an accurate registration scheme through intensive calibration using a vision-based object tracker. Their later work explored the potential of visuo-haptic AR technology for medical training (Harders et al. 2009). Ott et al. (2007) also applied an HMD-based visuo-haptic framework to training processes in industry. A similar setup was employed for cranial implant design in Scharver et al. (2004) and for an MR painting application in Sandor et al. (2007).

The final categories for *haptic mixed reality* (hMR), vR-hMR, vMR-hMR, and vV-hMR, lie in the middle column of the composite continuum. A common characteristic is that synthetic haptic signals generated from a haptic interface modulate or augment the haptic stimuli that occur due to a contact between a real object and the haptic interface tool. The VisHap system (Ye et al. 2003) can be an instance of vR-hMR that provides mixed haptic sensations in a real environment. In this system, some information about a virtual object (e.g., shape and stiffness) is generated by a haptic device, whereas other properties (e.g., texture and friction) are supplied by a real prop attached at the end-effector of the device. Other examples in this category include the SmartTool (Nojima et al. 2002) and SmartTouch systems

(Kajimoto et al. 2004). Using various sensors that capture real but small signals that can hardly be perceived by the bare hand, the authors translated the signals into haptic information and delivered them to the user to facilitate target tasks (e.g., peeling off the white from the yolk in an egg). The MicroTactus system (Yao et al. 2004) is another example, which detects and magnifies acceleration signals resulting from the interaction of a pen-type probe with a real object. This system was shown to improve the performance of tissue boundary detection in arthroscopic surgical training. A similar pen-type interface, Ubi-Pen (Kyung and Lee 2009), embedded miniaturized texture and vibrotactile displays in the pen, providing realistic tactile feedback for interaction with a touch screen. Other examples in this category are the FreeD milling system (Zoran and Paradiso 2013) and haptic augmentation on floor surfaces (Visell et al. 2009, see Chap. 3 for a review).

In contrast, environments in vV-hMR use synthetic visual stimuli. For example, Borst and Volz (2005) investigated the utility of haptic MR in a visual virtual environment by adding synthetic force to a passive haptic response for a panel control task. Their results showed that mixed force feedback improved task performance and user preference compared to pure synthetic force feedback. Haptically enhanced touch-panels, e.g., the TeslaTouch (Bau et al. 2010) is another example in this category. In vMR-hMR, both modalities rely on mixed stimuli. Ha et al. (2007) installed a vibrator in a real tangible prop to produce virtual vibrotactile sensations in addition to the real haptic information of the prop in a visually mixed environment, demonstrating that the virtual vibrotactile feedback enhances immersion for an AR-based handheld game. Bayart et al. (2007, 2008) introduced a teleoperation framework where force measured at the remote site is presented at the master site with additional virtual force and mixed imagery. In particular, they tried to modulate haptic stimuli such as contact forces with virtual feedback for a hole patching task and a painting application, which distinguishes the work from most of the other related studies.

Lastly, Bayart and Kheddar (2006) also suggested a simple taxonomy for haptic AR based on the functional aspect of a system. They termed a haptic AR application as "enhanced haptic" if haptic data from an information source was modulated or extrapolated in the application, e.g. providing active haptic guidance to sensorimotor skills (Lee and Choi 2010). In contrast, in applications for "haptic enhancing," fundamentally new information obtained from sources different from a haptic data source is added to the haptic data, e.g. haptizing non-haptic attributes such as weather variables on a geological map (Lee et al. 2008).

The next section gives an example of haptic mixed reality. We present a set of algorithms designed for modulation of real object stiffness by means of virtual force feedback.

12.3 Stiffness Augmentation

A person interacting with a real object constructs a mental image of it through perceiving distinct physical attributes of the object, such as shape, stiffness, friction, and surface texture. Consequently, haptic rendering of an object requires exclusive algorithms dedicated to each individual attribute as well as a combined framework that integrates these components. While this is a general notion valid in most haptic rendering systems, the same approach should also be followed in haptic AR. Accordingly, our recent research has focused on developing algorithms for augmenting individual haptic attributes, progressively moving towards a combined framework, a "haptic augmented reality toolkit."

Among various haptic attributes, our first focus was on the augmentation or modulation of real object stiffness, which is one of the most important attributes for perceiving shape and material. This section summarises a series of efforts that we made for stiffness modulation for more details, see Jeon and Choi (2008, 2009, 2010, 2011), Jeon and Harders (2012). The required system components, covering augmentation algorithms and hardware, are introduced. The presented frameworks are also the basis for an example application testbed outlined in Sect. 12.4.

The objective of stiffness augmentation is to provide a user with an augmented or modulated stiffness, by adding virtual force feedback when interacting with real objects. Two different kinds of interaction are examined in this section. The first system covered in Sect. 12.3.2 concerns stiffness modulation for single-point interaction. It supports typical exploratory patterns, such as tapping, stroking, or contour following. The second system, addressed in Sect. 12.3.3, extends the first one to two-point manipulation, focusing on grasping and squeezing interactions.

The emphasis of the algorithm design has been on minimising the need for prior knowledge and preprocessing for the geometric and material properties of real objects, while maintaining convincing perceptual rendering quality. In particular, our haptic AR systems require no geometry information of a real object. This allows us to avoid capturing the geometric model of a real environment, which otherwise would be time-consuming and require special equipment. Instead, we determine a-priori the dynamics model of a real object, employing the same interface as in the rendering stage. This strategy preserves a key advantage of AR; unlike in VR systems, a model of the entire environment is not required, which potentially leads to greater simplicity in the development of specific applications.

12.3.1 Rendering Hardware

As shown in Fig. 12.2, the haptic interface used in our haptic AR system consists of general impedance-type haptic interfaces (SensAble Technologies; PHANToM premium model 1.5), each of which has a custom-designed tool that is employed for interaction with a real object. The number of haptic interfaces depends on the number of contact points considered during manipulation. Stiffness augmentation in single-contact interaction (Sect. 12.3.2) requires one PHANToM, while the augmentation in two-contact squeezing needs two devices (Sect. 12.3.3). Each haptic tool is instrumented with a 3D force/torque sensor (ATI Industrial Automation, Inc.; model Nano17) attached between the tool tip and the gimbal joints at the last link of the

Fig. 12.2 Haptic AR hardware comprising two kinesthetic feedback devices. Reprinted, with permission, from Jeon and Harders (2012)

PHANToM. This configuration enables the system to measure the sum of two force components: force from the haptic interface and force from the user's hand.

12.3.2 Stiffness Modulation in Single-Contact Interaction

Our haptic AR systems target interaction with elastic objects of moderate stiffness. Objects made of plastic (e.g. clay), brittle (e.g. glass) or high stiffness material (e.g. steel) are not considered due to either complex material behaviour or the performance limitations of current haptic devices. In addition, we assume homogeneous dynamic material responses for the model-based estimation of real object deformation.

Indenting a real elastic object with a probing tool results in a deformation, while the user experiences a reaction force. We denote the apparent stiffness perceived by the user at time t by $k(t)$. Note that this is the stiffness when no additional virtual force is rendered. The goal of stiffness augmentation is now to alter the user-perceived stiffness, from $k(t)$ to a desired stiffness $\tilde{k}(t)$, by providing an adequate virtual force to the user's hand.

Let the force exerted at the tool tip through the haptic device be $\mathbf{f}_d(t)$ and that through the user's hand be $\mathbf{f}_h(t)$, as shown in Fig. 12.3a. The two force components deform the object surface and result in a reaction force $\mathbf{f}_r(t)$ in a steady state, such that

$$\mathbf{f}_r(t) = -\{\mathbf{f}_h(t) + \mathbf{f}_d(t)\}. \tag{12.1}$$

The reaction force $\mathbf{f}_r(t)$ during contact can be decomposed into two orthogonal force components, as shown in Fig. 12.3b:

$$\mathbf{f}_r(t) = \mathbf{f}_r^n(t) + \mathbf{f}_r^t(t), \tag{12.2}$$

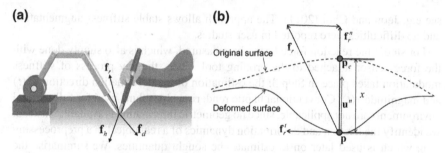

Fig. 12.3 Variable definitions for single-contact augmentation. Reprinted, with permission, from Jeon and Choi (2011). **a** For force computation. **b** For deformation estimation

where $\mathbf{f}_r^n(t)$ results from object elasticity in normal direction; and $\mathbf{f}_r^t(t)$ represents the tangential force, which is mainly due to friction between tool tip and object surface. Next, we consider the displacement $x(t)$ caused by the elastic force component. The magnitude describes the distance between the haptic interface tool position, $\mathbf{p}(t)$, and the original non-deformed position $\mathbf{p}_c(t)$ of a particle on the object surface. The unit vector in the direction of $\mathbf{f}_r^n(t)$ is denoted by $\mathbf{u}^n(t)$.

For a user to experience the target stiffness $\tilde{k}(t)$ at his or her hand, the augmentation should result in:

$$\tilde{\mathbf{f}}_h(t) = \tilde{k}(t)x(t)\mathbf{u}^n(t). \qquad (12.3)$$

Thus, using (12.1) the force that the haptic device needs to exert is

$$\tilde{\mathbf{f}}_d(t) = -\mathbf{f}_r(t) - \tilde{k}(t)x(t)\mathbf{u}^n(t). \qquad (12.4)$$

This equation indicates that in every haptic loop, the stiffness modulation algorithm requires four stpdf: (1) detection of the time instance at which the haptic tool touches the real object, (2) measurement of the reaction force $\mathbf{f}_r(t)$, (3) estimation of the direction $\mathbf{u}^n(t)$ and magnitude $x(t)$ of the resulting deformation vector, and (4) control of the device-rendered force $\mathbf{f}_d(t)$ to exert the desired force $\tilde{\mathbf{f}}_d(t)$. The following describes how these stpdf are addressed.

Step 1 is the contact detection. If the entire geometry of the real object is known, conventional haptic collision detection algorithms for virtual objects could be employed. However, this reduces the flexibility of the system since modelling would be required. Instead, we utilise force sensor readings for contact detection. A collision is assumed to have occurred when forces sensed during interaction exceed a small threshold. In this approach, the time delay between actual and detected contact depends on the force sensing quality and the threshold. We have developed algorithms to suppress noise, as well as to compensate weight and dynamics effects of the tool details are available in Jeon and Choi (2011). With this strategy, the time delay of contact detection was in most cases found to be smaller than 4 ms. This compares well with known perceptual thresholds of tactile simultaneity (20–30 ms) see e.g. Jeon and Choi (2011). The approach allows stable stiffness augmentation and no difficulties were reported in user studies.

For step 2 the reaction force has to be measured, which is also simply done with the force sensor attached to the probing tool. Then, the key process of stiffness modulation takes place in Step 3: the estimation of the deformation direction $\mathbf{u}^n(t)$ and magnitude $x(t)$. Conventional haptic rendering algorithms used for fully virtual environments are not applicable, since no geometric information is available. Instead, we identify the friction and deformation dynamics of a real object in a preprocessing step, which is used later on to estimate the sought quantities. We summarise the details of this process below.

Prior to actual rendering, two preprocessing stpdf have to be carried out for the object to be augmented. First, the friction between the real object and the tool tip is identified following Jeon and Choi (2011). For this a physical model-based approach is taken, employing the Dahl friction model. The original Dahl model is transformed into an equivalent discrete-time equation, as introduced in Mahvash and Okamura (2006). It is combined with a velocity-dependent term to cope with viscous friction. For the actual friction identification, we introduced an online technique that avoids complex nonlinearities. An adapted divide-and-conquer strategy is followed by performing the identification separately for the presliding and the sliding regime. This allows us to divide the nonlinear identification problem into two linear problems. Data are acquired during manual stroking by a user. The former include lateral displacement, velocity, normal force, and friction force, which are divided into two bins according to the lateral displacement. The data bin having smaller displacements (i.e. data in the presliding regime) is used to identify a parameter defining behaviour at low velocity; while the other is used for viscous and Coulomb parameters. Both processes use a linear recursive least-squares algorithm to determine the parameters.

Following this, the second preprocessing step is to identify the deformation dynamics of the real object. We rely on the Hunt-Crossley contact dynamics model (Hunt and Crossley 1975) to identify the nonlinear responses of an object. In the model, the response force magnitude under a certain displacement $x(t)$ and velocity $\dot{x}(t)$ is determined by

$$f(t) = k\{x(t)\}^m + b\{x(t)\}^m \dot{x}(t), \qquad (12.5)$$

where k and b are stiffness and damping constants, and m is a constant exponent (usually $1 \leq m \leq 2$).

For the identification, a user repeatedly presses and releases the elastic sample in normal direction, while data triples are recorded consisting of displacement, velocity, and reaction force magnitudes, along the surface normal. The recorded data are passed to a recursive least-squares algorithm for an iterative Hunt-Crossley model parameter estimation (Haddadi and Hashtrudi-Zaad 2008). In general, the overall identification process takes around 20–40 s.

Finally, in Step 4, based on the a-priori acquired information, the following computational process is carried out in every haptic rendering frame, in which the probing tool is detected as being in contact. The core part of the rendering is to estimate two

where $x_*(t)$ is the displacement of the deformation resulting from squeezing and $\mathbf{u}_*(t)$ is a unit vector representing the direction of that deformation. With (12.9) and (12.10), the virtual force that the haptic interface should provide to achieve the desired augmentation is

$$\tilde{\mathbf{f}}_{d,*}(t) = \mathbf{f}_{sqz,*}(t) - \tilde{k}(t)x_*(t)\mathbf{u}_*(t). \tag{12.11}$$

According to (12.11), during rendering the squeezing force $\mathbf{f}_{sqz,*}(t)$, the displacement $x_*(t)$, and direction $\mathbf{u}_*(t)$ of the deformation need to be estimated for each contact point, based on measurable variables $\mathbf{f}_{r,*}(t)$ and tool tip positions $\mathbf{p}_*(t)$.

The squeezing force $\mathbf{f}_{sqz,*}(t)$ cannot be directly calculated since $\mathbf{f}_{w,*}(t)$ is not known. Instead, we derive these forces according to the following three observations about an object held in a steady state. First, the two squeezing forces $\mathbf{f}_{sqz,1}(t)$ and $\mathbf{f}_{sqz,2}(t)$ share the same magnitude, but are in opposite direction ($\mathbf{f}_{sqz,1}(t) = -\mathbf{f}_{sqz,2}(t)$). Second, to hold the object stably without torques, each squeezing force should fall on the line connecting the two contact locations. Third, since the object is supported against gravity only by the two contacts, the sum of the two reaction force vectors is equal to the total weight of the object: $\mathbf{f}_{r,1}(t) + \mathbf{f}_{r,2}(t) = \mathbf{f}_{w,1}(t) + \mathbf{f}_{w,2}(t)$.

From our first and second observation, the directions of $\mathbf{f}_{sqz,*}(t)$ ($= \mathbf{u}_*(t)$) can be easily obtained using the line segment from $\mathbf{p}_1(t)$ to $\mathbf{p}_2(t)$ ($l(t)$ in Fig. 12.7). Its magnitude $f_{sqz,*}(t)$ is determined in a few stpdf. The total magnitude of the reaction forces along the squeezing direction is given by the magnitudes of the projections of the two reaction force vectors onto the direction of $l(t)$, $\mathbf{u}_l(t)$.

$$f_{r\downarrow sqz}(t) = |\mathbf{f}_{r,1}(t) \cdot \mathbf{u}_l(t)| + |\mathbf{f}_{r,2}(t) \cdot \mathbf{u}_l(t)|. \tag{12.12}$$

Note that $f_{r\downarrow sqz}(t)$ includes not only the two squeezing forces, but also the weight. The magnitude of only the squeezing force can be calculated by subtracting the effect of weight along $l(t)$ from $f_{r\downarrow sqz}(t)$:

$$f_{sqz}(t) = f_{r\downarrow sqz}(t) - f_{w\downarrow sqz}(t). \tag{12.13}$$

$f_{w\downarrow sqz}(t)$ is given according to the third observation by

$$f_{w\downarrow sqz}(t) = \left|\left(\mathbf{f}_{r,1}(t) + \mathbf{f}_{r,1}(t)\right) \cdot \mathbf{u}_l(t)\right|. \tag{12.14}$$

Finally, according to the first observation, the force magnitude at each contact point is

$$f_{sqz,1}(t) = f_{sqz,2}(t) = 0.5 f_{sqz}(t). \tag{12.15}$$

The next step is the estimation of the displacement $x_*(t)$ of the deformation due to squeezing in (12.11). Let d_0 be the distance between the contact points on the original non-deformed surface, which is constant over time due to our no-slip assumption.

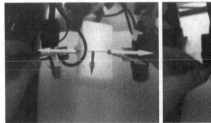

Fig. 12.5 Visuo-haptic augmentation. Reaction force (*red*), weight (*blue*), and haptic device force (*green*) are depicted. Examples with increased stiffness (virtual forces oppose squeezing) and decreased stiffness (virtual forces assist squeezing) are shown on *left* and *right*, respectively. Reprinted, with permission, from Jeon and Harders (2012)

d_0 is given by the two initial contact points, $\mathbf{p}_{c,1}(t)$ and $\mathbf{p}_{c,2}(t)$ in Fig. 12.7. With our initial assumption of a homogeneous object, we can derive $x_1(t) = x_2(t)$, and the displacements can be calculated by

$$x_1(t) = x_2(t) = 0.5\Big(d_0 - d(t)\Big), \tag{12.16}$$

where $d(t)$ is the distance between the points during the interaction. Here, we should note that the estimates of deformation direction and magnitude may differ from the actual quantities. Nevertheless, this presumably will have only little influence on judging stiffness since stiffness perception in squeezing is mainly based on relative distance between contact points and squeezing force magnitude. Finally, with these required variables, the desired force command for stiffness modulation, $\tilde{\mathbf{f}}_{d,*}(t)$, can be calculated according to (12.11).

In Jeon and Harders (2012), the physical performance and the result of a psychophysical experiment are reported. Overall, the evaluation indicates that our two-point stiffness modulation algorithm is also capable of augmenting object stiffness at a reasonable quality. Augmentation was possible regardless of the real base stiffness and direction of stiffness change.

The developed haptic augmentation system has further been merged with a visual AR framework (Harders et al. 2009). The visual system allows, for instance, the display of information related to the haptic augmentation, such as the force vectors involved in the algorithm. Example snapshots are depicted in Fig. 12.5.

12.4 Example Application: Augmentation of Stiffer Inclusions

In this section an application example will be outlined, in which virtual stiffer inclusions (i.e. simulated tumour tissue) are created inside real deformable sample objects, with the target of medical training (see Fig. 12.6). The general concept of the stiffness

Fig. 12.6 Setup used for haptic augmentation of stiffer tumor. Reprinted, with permission, from Jeon et al. (2012)

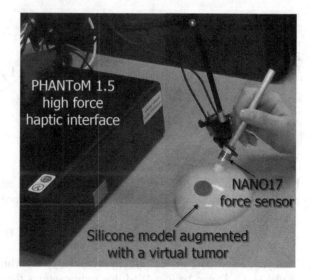

PHANToM 1.5
high force
haptic interface

NANO17
force sensor

Silicone model augmented
with a virtual tumor

augmentation remains the same as introduced in the previous sections. The goal is to alter the overall force $\mathbf{f}_H(t)$ experienced by the user, by rendering an additional virtual stimuli $\mathbf{f}_T(t)$ representing the tum-or, together with the real response $\mathbf{f}_R(t)$ of the object, at time t. We rely on nonlinear Hunt-Crossley models (Hunt and Crossley 1975) to augment the dynamic behaviour of a simulated embedded stiffer inclusion. Further details on the overall approach are available in Jeon et al. (2012). In a first step, appropriate parameters for the models have to be identified to represent the forces contributing to the overall feedback.

The underlying idea of identifying parameters to describe response forces of an embedded stiffer inclusion is to use two physical silicone hemispheres— one without and one with an enclosed physical nodule. Estimating the difference between the two samples allows to determine the contribution of only the inclusion itself. To this end, the same haptic acquisition hardware as described above is employed to obtain interaction data when probing both samples. Positions are acquired through haptic device encoders, while contact velocities are estimated using a first-order adaptive windowing filter (Janabi-Sharifi et al. 2000). The indentation forces are measured with the attached Nano 17 force/torque sensor. These data are employed to obtain an explicit-form Hunt-Crossley model $f = H(x, \dot{x})$ via a recursive least-squares algorithm.

First the force response model $f = H_{NT}(x, \dot{x})$ of the sample with no inclusion is obtained. Thereafter, interaction data for the model with an embedded stiffer nodule are acquired. The surface is contacted at point $\mathbf{p}_{T,s}$, closest to the centre $\mathbf{p}_T$ of the spherical inclusion. The indentation is performed towards $\mathbf{p}_T$. Figure 12.7 illustrates the locations of these two points. The collected data are denoted by $(x_{TE}, \dot{x}_{TE}, f_{TE})$.

With regard to force augmentation, $H_{NT}(x, \dot{x})$ represents the magnitude of $\mathbf{f}_R(t)$. Moreover, the forces f_{TE} measured from the model with inclusion combine $\mathbf{f}_R(t)$ and

Fig. 12.7 Quantities required in rendering algorithm for augmentaion of stiffer inclusion. Reprinted, with permission, from Jeon et al. (2012)

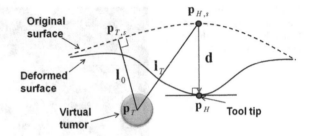

$\mathbf{f}_T(t)$. Thus, in order to obtain the magnitude of $\mathbf{f}_T(t)$, all data pairs $(x_{TE}, \dot{x}_{TE})$ are passed to $H_{NT}(x, \dot{x})$, and differences are computed according to

$$f_T(x_{TE}, \dot{x}_{TE}) = f_{TE} - H_{NT}(x_{TE}, x_{TE}^\cdot). \tag{12.17}$$

This yields new data triples $(x_{TE}, \dot{x}_{TE}, f_T)$, based on which another Hunt-Crossely model $H_T(x, \dot{x})$ can be identified that only captures the contribution of the stiffer inclusion. Note that this model specifies the force response at contact point $\mathbf{p}_{T,s}$, when probing into the direction of $\mathbf{p}_T$. In the following we will describe how the model is used to determine arbitrary feedback forces.

Based on the identified models, a heuristic rendering algorithm is employed to compute haptic feedback $\mathbf{f}_T(t)$ representing the tumour contribution during real-time augmentation. The same hardware as during the recording is employed. Figure 12.7 illustrates all variables involved in this process.

When the user contacts the sample with the probing tool at $\mathbf{p}_H(t)$ at time t, then the augmentation response force $\mathbf{f}_T(t)$ is given as

$$\mathbf{f}_T(t) = f_T(t) \frac{\mathbf{p}_H(t) - \mathbf{p}_T}{|\mathbf{p}_H(t) - \mathbf{p}_T|}. \tag{12.18}$$

The force $\mathbf{f}_T(t)$ is directed from inclusion position $\mathbf{p}_T$ to tool position $\mathbf{p}_H(t)$, and has a magnitude $f_T(t)$. Therefore, $f_T(t)$ has to be estimated based on the identified model.

The closest point from $\mathbf{p}_H(t)$ on the non-deformed sample surface is denoted as $\mathbf{p}_{H,s}(t)$. The displacement of this point due to deformation is $\mathbf{d}(t)$. The line segment between $\mathbf{p}_{H,s}(t)$ and $\mathbf{p}_T$ is denoted as $\mathbf{l}_T(t)$. Projecting the displacement vector $\mathbf{d}(t)$ onto $\mathbf{l}_T(t)$ yields a vector of length

$$x_{lt}(t) = \mathbf{d}(t) \cdot \frac{\mathbf{l}_T(t)}{|\mathbf{l}_T(t)|}, \tag{12.19}$$

where distance $x_{lt}(t)$ represents the displacement $\mathbf{d}(t)$ of $\mathbf{p}_{H,s}(t)$ into the direction of the inclusion.

Based on $x_{lt}(t)$, $f_T(t)$ is approximated using the previously obtained Hunt-Crossley model. Assuming material homogeneity, the response along $\mathbf{l}_T(t)$ can be approximated by $H_T(x, \dot{x})$. This requires the length of $\mathbf{l}_T(t)$ to be identical to that of $\mathbf{l}_0$. However, in general, $|\mathbf{l}_T(t)| \geq |\mathbf{l}_0|$. Therefore, we apply a simple scaling:

$$x(t) = x_{lt}(t)\frac{|\mathbf{l}_0|}{|\mathbf{l}_T(t)|}. \tag{12.20}$$

The distance $x(t)$ is a linearly-normalised deformation magnitude in relation to the reference deformation along $\mathbf{l}_0$. Based on this, the force magnitude due to the presence of the stiffer inclusion is estimated as

$$f_T(t) = H_T(x(t), \dot{x}(t)), \tag{12.21}$$

where $\dot{x}(t)$ is derived from $x(t)$ using a first-order adaptive windowing filter.

It should be noted that this algorithm is only heuristic. As before, it has been designed for real-time rendering, without any need of geometric information of the real objects. Various experiments with the algorithm have been conducted, as described in Jeon et al. (2012). The results demonstrated that the augmentation technique was capable of providing reasonable feedback approximating a stiffer inclusion in a soft sample object.

12.5 Conclusion and Discussion

In this chapter the new paradigm of haptic augmented reality has been overviewed. A taxonomy of the domain has been provided, extending Milgram's reality-virtuality continuum. Heuristic algorithms for stiffness augmentation have been outlined—considering contact at one or two surface locations. Finally, an example of stiffness augmentation in an applied context has been provided. A system ultimately targeting medical training of palpation has been described. Stiffer virtual inclusions in real deformable samples could be rendered, based on a modification of the previously introduced algorithms.

While the discussed stiffness augmentation algorithms constitutes a first step towards the envisioned haptic augmented reality toolkit, there are still numerous clear limitations and assumptions. All interactions are tool-mediated. It would be more appropriate to allow direct finger interactions, however, this necessitates the development of more advanced hardware, including miniaturised force sensors. Furthermore, tissue homogeneity is assumed for the real mock-ups in the presented examples. In addition, only very basic shapes were employed for the test objects. It remains to be seen if the proposed dynamics model can also be extended to inhomogeneous contact dynamics. Moreover, the algorithms mainly consider indentations only in normal direction. Tangential probing motions should also be allowed. In future work, further extensions in these domains are necessary to develop practical applications employing haptic augmented reality. Nevertheless, the presented algorithm lay the groundwork for such endeavours.

Acknowledgments The described research has been supported in parts by Korean government programs, NRL R0A-2008-000-20087-0 from NRF and ITRC (NIPA-2011-)C1090-1111-0008 from NIPA, the Korean-Swiss Science and Technology Cooperation program 2010/2011, and the EU project BEAMING 248620.

References

Bau O, Poupyrev I, Israr A, Harrison C (2010) Teslatouch: electrovibration for touch surfaces. In: Proceedings of the 23nd annual ACM symposium on user interface software and technology, UIST '10, pp 283–292

Bayart B, Didier JY, Kheddar A (2008) Force feedback virtual painting on real objects: a paradigm of augmented reality haptics. Lect Notes Comput Sci (EuroHaptics 2008) 5024:776–785

Bayart B, Drif A, Kheddar A, Didier JY (2007) Visuo-haptic blending applied to a tele-touch-diagnosis application. Lect Notes Comput Sci (Eurohaptics 2007) 4563:617–626

Bayart B, Kheddar A (2006) Haptic augmented reality taxonomy: haptic enhancing and enhanced haptics. In: Proceedings of EuroHaptics, pp 641–644

Bennett E, Stevens B (2006) The effect that the visual and haptic problems associated with touching a projection augmented model have on object-presence. Presence: Teleoperators Virtual Environ 15(4):419–437

Bianchi G, Jung C, Knörlein B, Székely G, Harders M (2006a) High-fidelity visuo-haptic interaction with virtual objects in multi-modal AR systems. In: Proceedings of the IEEE and ACM international symposium on mixed and augmented reality, pp 187–196

Bianchi G, Knörlein B, Székely G, Harders M (2006b) High precision augmented reality haptics. In: Proceedings of EuroHaptics, pp 169–168

Billinghurst M, Kato H, Poupyrev I (2001) The magic book-moving seamlessly between reality and virtuality. IEEE Comput Graph Appl 21(3):6–8

Borst CW, Volz RA (2005) Evaluation of a haptic mixed reality system for interactions with a virtual control panel. Presence: Teleoperators Virtual Environ 14(6):677–696

Ha T, Chang Y, Woo W (2007) Usability test of immersion for augmented reality based product design. Lect Notes Comput Sci (Edutainment 2007) 4469:152–161

Haddadi A, Hashtrudi-Zaad K (2008) A new method for online parameter estimation of Hunt-Crossley environment dynamic models. In: Proceedings of the IEEE international conference on intelligent robots and systems, pp 981–986

Harders M, Bianchi G, Knörlein B, Székely G (2009) Calibration, registration, and synchronization for high precision augmented reality haptics. IEEE Trans Vis Comput Graph 15(1):138–149

Hunt K, Crossley F (1975) Coefficient of restitution interpreted as damping in vibroimpact. ASME J Appl Mech 42:440–445

Janabi-Sharifi F, Hayward V, Chen CSJ (2000) Discrete-time adaptive windowing for velocity estimation. IEEE Trans Control Syst Technol 8(6):1003–1009

Jeon S, Choi S (2008) Modulating real object stiffness for haptic augmented reality. Lect Notes Comput Sci (EuroHaptics 2008) 5024:609–618

Jeon S, Choi S (2009) Haptic augmented reality: taxonomy and an example of stiffness modulation. Presence: Teleoperators Virtual Environ 18(5):387–408

Jeon S, Choi S (2010) Stiffness modulation for haptic augmented reality: extension to 3D interaction. In: Proceedings of the haptics symposium, pp 273–280

Jeon S, Choi S (2011) Real stiffness augmentation for haptic augmented reality. Presence: Teleoperators Virtual Environ 20(4):337–370

Jeon S, Choi S, Harders M (2012) Rendering virtual tumors in real tissue mock-ups using haptic augmented reality. IEEE Trans Haptics 5(1):77–84

Jeon S, Harders M (2012) Extending haptic augmented reality: modulating stiffness during two-point squeezing. In: Proceedings of the haptics symposium, pp 141–146

Kajimoto H, Kawakami N, Tachi S, Inami M (2004) SmartTouch: electric skin to touch the untouchable. IEEE Comput Graph Appl 24(1):36–43

Kyung KU, Lee JY (2009) Ubi-Pen: a haptic interface with texture and vibrotactile display. IEEE Comput Graph Appl 29(1):24–32

Lee C, Adelstein BD, Choi S (2008) Haptic weather. In: Proceedings of the symposium on haptic interfaces for virtual environments and teleoperator systems, pp 473–474

Lee J, Choi S (2010) Effects of haptic guidance and disturbance on motor learning: potential advantage of haptic disturbance. In: Proceedings of the IEEE haptics symposium (HS), pp 335–342

Mahvash M, Okamura AM (2006) Friction compensation for a force-feedback telerobotic system. In: Proceedings of the IEEE international conference on robotics and automation, pp 3268–3273

Milgram P, Colquhoun H Jr (1999) A taxonomy of real and virtual world display integration. In: Tamura Y (ed) Mixed reality-merging real and virtual worlds. Springer, Berlin, pp 1–16

Nojima T, Sekiguchi D, Inami M, Tachi S (2002) The SmartTool: a system for augmented reality of haptics. In Proceedings of the IEEE virtual reality conference, pp 67–72

Ott R, Thalmann D, Vexo F (2007) Haptic feedback in mixed-reality environment. Vis Comput: Int J Comput Graph 23(9):843–849

Sandor C, Uchiyama S, Yamamoto H (2007) Visuo-haptic systems: half-mirrors considered harmful. In: Proceedings of the world haptics conference, pp 292–297

Scharver C, Evenhouse R, Johnson A, Leigh J (2004) Designing cranial implants in a haptic augmented reality environment. Commun ACM 47(8):32–38

Vallino JR, Brown CM (1999) Haptics in augmented reality. In: Proceedings of the IEEE international conference on multimedia computing and systems, pp 195–200

Visell Y, Law A, Cooperstock J (2009) Touch is everywhere: floor surfaces as ambient haptic interfaces. IEEE Trans Haptics 2(3):148–159

Yao H-Y, Hayward V, Ellis RE (2004) A tactile magnification instrument for minimally invasive surgery. Lect Notes Comput Sci (MICCAI) 3217:89–96

Ye G, Corso J, Hager G, Okamura A (2003) VisHap: augmented reality combining haptics and vision. In: Proceedings of the IEEE international conference on systems, man and cybernetics, pp 3425–3431

Printed in the United States
By Bookmasters

Printed in the United States
By Bookmasters